DANIEL DRAWBAUGH

The Edison of the Cumberland Valley

DANIEL DRAWBAUGH

The Edison of the Cumberland Valley

by

WARREN J. HARDER

Philadelphia
University of Pennsylvania Press

Published in Great Britain, India, and Pakistan
by the Oxford University Press
London, Bombay, and Karachi

Library of Congress Catalogue Card Number: 59-9201

To my Godson—
Jeffery Warren Baker
To whom the telephone now
is but a tinkling terror.

Printed in U. S. A.

Contents

Illustrations

c. the "old curiosity shop"
d. the estate acquired from Sara J. Haly
e. the home built by Drawbaugh for his daughter Laura
f. deed showing sale by Drawbaugh of former Haly property
g. the Cumberland Road residence

DANIEL DRAWBAUGH

The Edison of the Cumberland Valley

Introduction

Many were the legal battles involving the telephone—some small and easily disposed of, others large and troublesome. But, in all likelihood, none ever caused the great Bell system more headaches than one centering around a poor country inventor —Daniel Drawbaugh, the so-called "Edison of the Cumberland Valley."

Most certainly, none of the early telephone litigations came closer to changing the course of "talking machine" history than the action of the American Bell Telephone Company versus the People's Telephone Company in the 1880's. Prime figure in this eight-year litigation was Daniel Drawbaugh. But for a single vote in the highest court of the land, we might commonly associate the telephone with Drawbaugh instead of Bell.

Daniel Drawbaugh was poor—born poor and died poor. In between, he led a most eventful life, and it is not entirely outside the realm of possibility to speculate that his lack of ready money prevented him from obtaining the important patents which could have made major differences in the telephone systems of our great nation.

No one can say with authority that the "Wizard of Eberly's Mills," did antedate Professor Bell. The Supreme Court, by a one-vote majority, said he did not, and who should argue with the High Court? The fact remains that this comparatively uneducated, central Pennsylvania country mechanic, inventor, and patentee did provide the basis for giving the Bell organization some troublesome moments—days, weeks, months, and even years.

Daniel's work was not limited to "talking machines"— although telephones occupied some of his most prized hours. His inventions and improvements were spread over a wide range, and he was able to sell many.

The Telephone of 1664.

And as glasses have highly promoted our seeing, so 'tis not improbable, but that there may be found many mechanical inventors to improve our other senses, of hearing, smelling, tasting, touching. 'Tis not impossible to hear a whisper a furlong's distance, it having been already done; and perhaps the nature of the thing would not make it more impossible, though that furlong should be ten times multiplied. And though some famous authors have affirmed it impossible to hear through the thinnest plate of Muscovy glass, yet I know a way by which it is easy enough to hear one speak through a wall a yard thick. It has not yet been thoroughly examined how far Otocousticons may be improved, nor what other ways there may be of quickening our hearing, or conveying sound through other bodies than the air; for that is not the only medium. I can assure the reader that I have, by the help of a distended wire, propagated the sound to a very considerable distance in an instant, or with as seemingly quick a motion as that of light, at least, incomparably swifter than that, which at the same time was propagated through the air; and this not only in a straight line, or direct, but in one bended in many angles.—*From works of Robert Hooke, published in* 1664.

1. A quotation from Robert Hooke.

2. He, too, knew a way.

This photograph of Daniel Drawbaugh was made, probably in the late winter of 1881-82, when he was in New York City for the instruments tests. Photographer Moreno had his studio in the same building as the offices of The People's Telephone Company.

However, it was not an easy matter for Daniel to come into money, although he made an auspicious beginning in this direction. A biographical sketch compiled by Cumberland County attorney-historian Jeremiah Zeamer said: "While yet a mere boy he earned considerable spending money by making boot trees and other articles of actual use . . . " Money was not considerable, however, through most of his years.

Apparently if Zeamer had been the judge, Daniel would have come into wealth through the telephone. In 1905, he said flatly that Daniel had invented a telephone in 1867, some time before Bell's first patent in 1876. Zeamer also said: "Although not successful in securing a patent for same, Daniel Drawbaugh is none the less the original inventor of the telephone."

Then loved by almost all who knew him, today still held in esteem by people throughout the Drawbaugh home region, the fact remains that Daniel did not have the votes of a controlling number of Supreme Court justices—and the lack of those votes spelled out a result so very typical of his life—disappointment.

Called a genius by some, a good mechanic and imitator by others of his day, it was generally acknowledged that he was a poor businessman—and that had much to do with the course of his life. Affiliated with million-dollar corporations, he realized little from most ventures and nothing from others. Throughout his life he owed money.

On the rare occasions that he came into a little capital, it would be depleted rapidly—paying past due bills. Once he received five hundred dollars for an interest in an invention; he squared some old accounts and within a week had to start borrowing again.

Not too long after that experience, it was necessary to borrow five dollars to attend his father's funeral, but the family could not go for lack of additional funds—and suitable clothing.

The Civil War cost him money. It took many years for the country to recover from the war, but Daniel never did. A profitable business was ruined near the start of the conflict when a southern agent pocketed money due Daniel. On another occasion, he sold an interest in a patent right for a worthless note on which he borrowed. Later, he had to make restitution.

Despite his trials and troubles, he was able to find a thousand dollars to apply toward the purchase of a farm for his aged and infirm parents, and somehow obtained an additional four hundred dollars to buy them furniture, clothing, etc.

His kindness wasn't limited to his parents. His workshop was a congregating place for the village, and he never was too busy to take time to demonstrate his new devices. His kindness also extended to youth and some of his best moments of delight came in fixing things for the young folks. Especially popular with the older boys was his electrical "shocking machine." Today, the value of "electric shock" treatment is recognized—the youth of Eberly's Mills knew it over three-quarters of a century ago.

The inventor was a man of contradictions. There was no time to exterminate the field mice that overran the shop or to screen the windows against flies, but to devise and install a fly chaser on the shop ceiling—yes.

Unlike most inventors, he enjoyed having people about him and openly discussed his devices. He was, however, most sensitive about his "talking machines" and sometimes reverted to secrecy when fearful of ridicule. When he once spoke of conversing over a wire "across the ocean" he was told by the village undertaker, William Darr, to try it first across the Yellow Breeches Creek. Daniel never forgave him.

His early 'phone was tested secretly at night on a wire fence by picking up the sound of a ticking clock in the shop. On one occasion, he walked to White Hill in the dead of night to test a newly made sounder on the telegraph line there. A wire was thrown over the line, connected to the instrument, and—it worked.

His day and night labors might lead one to conclude that he had no interest in his family. Such was not the case; there were close family ties even in spite of a nagging wife who went to the extent of breaking his photographic apparatus so that no more time and money would be wasted on it. Although he couldn't afford it, he bought an organ for his family and had one of the shop workmen give music lessons to his "Emmy"—

the daughter who had helped him with the "talking machine" experiments.

Misfortune seemed to plague Daniel constantly. Of the eleven children, some were sickly, and he always was in debt to the doctor. Records indicate that only four offspring survived him.

But, even with their poverty, the Drawbaughs were proud people. Daniel forever stood by his beliefs, even sometimes when confronted with sounder reasoning, and, to his dying day, claimed his telephone secret was stolen from him. His wife, Elcetta, who survived him by two years, also was proud. When she detected grey hair, she insisted that daughter Belle dye it for her immediately.

Although he never made the money he should have, Daniel did prove he was much more than a mediocre country mechanic. A total of seventy patents was issued to him, and he produced hundreds of devices for himself and others for which no protection was obtained. It is interesting to observe that, although denied his claim as originator, of his seventy patents thirty-five pertained to telephones. A bill introduced in the first session of the Fifty-fourth Congress, to grant him certain patents, had two readings and died in committee.

Originator or imitator, genius or charlatan, believe what you will; the "wizard" still owns the distinction of almost bringing the American Bell Telephone Company to its knees.

The author felt that he was as well prepared to write of Daniel Drawbaugh, as were some of the witnesses to testify. He could, like them, lay claim to knowledge of his subject on several counts. After all, he

(1) visited Eberly's Mills as a boy on many a Sunday, and played hooky from church (while his father taught the men's Bible class) to explore the Drawbaugh scene;

(2) attained the distinction, and the close association which came with a fall through the rotted floor of the "wizard's" then-abandoned shop;

(3) ate innumerable Sunday dinners at the table of Mabel Hart Drawbaugh, daughter-in-law of the subject of this theme; and there, assimilated information along with his dessert; and all of these things, more years ago than he likes to remember;

(4) sold real estate for an office located, in recent years, at 813 North Second Street, Harrisburg, in the former home of attorney Michael W. Jacobs, who played his part in shaping the destiny of Daniel Drawbaugh.

But, faced with the stern necessity of producing what he hopes is a factual account, he found his knowledge as woefully lacking as did some of the witnesses, theirs. Had he years ago the eyes to see and the ears to hear the many things ignored, this writing might have been an easier task.

Through the past decade, as his files of historical material grew, Drawbaugh data were added against the day that inspiration would overcome good judgment and the mountain labor and bring forth . . .?

And too, there were printed records of the litigation, newspaper files, and folks living who knew our subject and his work. To these sources, he went seeking information. He found, among other things, a unanimity of opinion sometimes expressed vehemently but always with finality on the subject of who had invented the telephone.

He gratefully acknowledges a debt of gratitude to:

Ethel, who twenty-five years ago said, "Let's write a book"; to Lillie Haas, who knew people; to Mabel Schultz, who kept clippings and memories; to Mabel Hart Drawbaugh, whose contribution already has been acknowledged; to Clarence Fettrow, hale and hearty son of blacksmith Daniel Fettrow of our story, and who provided valuable clews; to Ellis Eichelberger, schoolmate of the Drawbaugh children, who preserved photographs and memories; to Billy Bowman, who is mentioned in the text; to Willard Smiley, who once owned one of the inventor's clocks; to Caroline Spangenberg Bucher, who has accumulated some material of interest; to "Marty" Brinton, venerable real estate broker, whose memory spanned the past half-century with ease; to Naomi Arnold Irvine, whose collection of historical books is amazing; to Stewart Otstot, who tore his store apart to locate a photograph; to Rankin Sutton, who told what he knew; to H. J. Smith, who keeps old indentures; to Richard Bair, machinist and inventor, whose shop stands but a few feet from the spot where was located the shop of our subject, and who knew of his patents; to Mrs. George May, friend and neighbor of the Drawbaughs away back when, and whose memory had not dimmed in spite of her ninety-eight years; to John Drawbaugh, whose grandfather was the inventor's brother, and who made his contribution; to Margaret Oyster Bream, granddaughter of Daniel Drawbaugh, and who supplied valuable information; to Congressman Walter M. Mumma, who, in the writer's opinion, certainly earned his reelection; to the girls of the state library and the law library staffs; to Greta Mullen, former curator of the Dauphin County Historical Society, to whom nothing was too much trouble; to attorney Bob Derrick, who, it seems, can explain almost any legal procedure by the application of a watermelon and shotgun analogy; to Bob Kamensky, manuscript editor, who in turn had to be edited at times; to his Mabel, a most efficient and able typist; and to all of these kind hearts and gentle souls whose efforts, individually and collectively, made possible this writing of the story of Daniel Drawbaugh, he sincerely tenders his thanks.

I

The Early Years

About 1820, blacksmith John Drawbaugh moved to Allen Township in Cumberland County, Pennsylvania, to rear his family and ply his trade. He settled at a spot not far from the conflux of Cedar Run and Yellow Breeches Creek. The home site overlooked a few cabins, a gristmill, and a sawmill built in the 1760's on land of the manor of Louther. There, in a log house, on July 14, 1827, his fourth child, Daniel, was born.

The boy inherited a strong mechanical aptitude which, at an early age, was turned to good use in the production of useful things. In country school, his progress was just average; the father's shop held more interest. There, his urge to make things could be satisfied. He often told of making parts for a clock when he was just ten or twelve years old. A few years later, he made a small steam engine which was used to power the shop grindstone, and a machine to saw the segments of the rim of a wagon wheel. At seventeen, he made a fine rifle for Daniel Balsley, for which he received a fancy price. A farmer of the neighborhood needed a mowing machine—soon after, young Daniel had it completed.

The Allen Township area had natural attractions. The good limestone land induced farmers of means to locate there. Nicholas Kisecker, Jr., under partition proceedings of July 30, 1805, obtained a parcel of land along Cedar Run, upon which he erected a fine stone gristmill. The old mill property then was converted into living quarters. About 1842, a church was established on the hill adjoining what is said to have been the first Drawbaugh home and smith shop. With its gristmill, sawmill, and oilmill in an eighth-mile radius, the settlement logically was called Milltown.

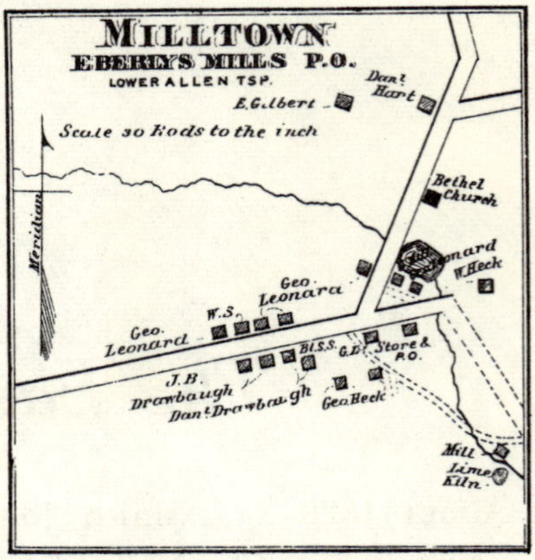

3. It all began here.

4. Once a log house stood on this site—
the inventor's birthplace—so it is said.

5. Milltown's first gristmill was the stone building shown on the right. It stood at Cedar Run and the village spring. Erected in the 1760's. Owned by Edward Ward in 1769. Acquired by Nicholas Kisecker, Jr. in 1805. Remodelled into living quarters about 1810, it was the home of the mill proprietors during the years following. The large frame addition subsequently was added. In 1869, Henry Yetter's tavern occupied it. From 1873-76, it was tenanted by four families including Holsinger, the village printer who also operated his shop there. Demolished prior to 1915.

Historical writings make but scant mention of the first gristmill and a clovermill at Yellow Breeches Creek, except to indicate that the latter, during one period, might have "fallen from grace" and operated as a distillery.

The Cumberland Valley Railroad built its tracks near the Harrisburg Turnpike, about a mile away. A stop was established at "Whitehill's" (named for Robert Whitehill who had located there about 1772) and the community had train service from 1837.

The Hertzlers acquired extensive land holdings along the Lisburn Road and there established their large farm with orchard and winery. This family also owned another farm in Milltown proper, and a third at the edge of the settlement.

Another wealthy landowner was the Gorgas family. William R. Gorgas was born there on his father's farm in 1806. The old homestead still stands, with little exterior change, at the Carlisle and Lisburn roads. He served as representative in the Pennsylvania State Assembly from 1836 to 1839, and as a state senator from 1842 to 1844. The family owned and operated the Gorgas' mine which produced iron ore for a Harrisburg furnace.

In 1844, Christian Eberly located at Milltown, later purchased the old Kisecker mill, and was instrumental in having the post office re-established. For postal purposes, the community was named officially Eberly's Mills and bears that name today.

Allen Township was partitioned in 1850 and the area with which we are concerned became Lower Allen Township.

In 1856, two years after his marriage to Elcetta Thompson, Daniel Drawbaugh moved from his parents' home into a large, barnlike house back of the gristmill. The basement (ground floor) was outfitted as a workshop.

Landlord, postmaster, mill operator, Christian Eberly, who also owned most of the houses in the settlement, lived up to his given name. He bought a few simple hand tools and a hand-powered lathe for the potential inventor. It was understood he was to be repaid by a half interest in any new inventions which might be forthcoming.

Daniel got off to what appeared to be a successful and promising start, but, as misfortune plagued him throughout his life, his prosperity was not a lasting matter. At the time he moved, he owned two patents. The first he obtained was for improvements on a stave jointing machine (No. 8,505 issued November 11, 1851) and the second was for a stave machine (No. 12,900 issued May 22, 1855). These devices became especially popular in the South, and Daniel's agent there sold many with exclusive territories for their use.

With the outbreak of the Civil War imminent, the salesman stopped selling and began collecting. With a pocketfull of cash, notes, and promises, he forgot the home office at Eberly's Mills and marched away under the Stars and Bars. Drawbaugh estimated his loss at between $4,000 and $5,000.

The manufacture of the machines was continued in the basement workshop until the end of 1867, but profits were little. Some years later, Daniel reflected: "It didn't amount to three or four hundred dollars a year for a number of years."

The inventor obtained additional income from model building, repairs to agricultural implements and other mechanical devices. In 1863, he received a patent for improvements in mill stones, and another for an improved machine to level the faces of the stones. A machine for feeding nail plates was in the model stage, and patent No. 51,435 was issued on it December 12, 1865. An improved nail plate feeding device patent followed in 1867.

The bitter cold winter of 1865-66 proved a blessing for Daniel although village storekeeper, Lafayette Croll, had his troubles because of it. His customers wanted molasses on their buckwheat cakes—but the molasses wouldn't come from the barrel. The despairing Croll sent word to Daniel. He came, conferred, and returned the next day with a crude, wooden measuring faucet and the laconic invitation, "Try it." That day, cold as it was, Eberly's Mills had molasses.

The canny storekeeper always had a little something laid back for a good thing. A deal was made and Daniel had another backer—Croll would finance the patent for a share. Drawbaugh perfected the device and the following November secured two patents on it. A. H. Gardner, a neighbor with some cash and a proclivity for signing notes, saw it and wanted a share. The inventor sold part of his interest for a note; Croll sold all of his for cash, and faucetless, forswore storekeeping to become a house painter in Harrisburg.

In May, 1867, a group of men organized under the name, The Drawbaugh Pump Company, to manufacture and market rotary pumps—an improved version of the faucet. A consolidation of faucet and pump interests was effected, and A. H. Gardner, Samuel M. Hertzler, William L. Gorgas, Joseph F. and John M. Hursh, Henry S. Rupp, John Sheely, William R. Gorgas, and Daniel Drawbaugh were ready to manufacture the patents. The old Eberly Clovermill along the Yellow Breeches Creek was leased for a shop, and, with the close of the year, remodeling was nearly completed.

This three and a half story, stone and frame structure, which had last been used as a match factory, had ample space and was equipped with water power. In early 1868, operations were underway in the new quarters. Daniel was given a workshop there which he continued to use until 1911.

Somewhat of a failure developed in another venture, and he lost some of his supporters. The model nail plate feeding machine, standing in a corner of his shop, failed to receive more than casual attention until Eberly mentioned it to a friend, Frank Lee, who was secretary to, and brother-in-law of, Pennsylvania's Governor Geary (1867-72). They viewed the machine and bargained for a share.

At the sixteenth annual fair of the Pennsylvania State Agricultural Society, held at Harrisburg from Tuesday, September 29 until Friday, October 2, 1868, Daniel exhibited a steam pump and the nail feeder. Perhaps the pump was beyond the ken of the judges, for they ignored it, but for the feeder voted him a "diploma," as a newspaper of the day put it.

Installed for trial in the McCormick Nail Mill at West Fairview, it failed to supply the answer to nail makers' problems and never was marketed. About 1873, the rosy bubble burst—Eberly, Geary, and Lee knew that they had made a bad investment. Lee later filed a civil suit against Drawbaugh for thirty-four dollars in an attempt to recover something. The money was paid, and politics and inventive genius parted company.

6. The William R. Gorgas messuage—iron ore paid off—faucets cost four grand.

7. Not "16 tons"—but 40,000 of choice iron ore were taken from here long ago for the Porter Furnace at Harrisburg. Today, a picnic area.

8. The birthplace of Drawbaugh's "talking machine" was the basement workshop of this house behind the gristmill. Here, from 1856 until the end of 1867, he manufactured his stave jointing machines, nail plate feeder and first measuring faucet. The structure was destroyed by fire early in 1939. (Sketched by Pat Thornton from drawings and descriptions of Clarence Fettrow and Richard and Horace Bair for the author's collection of historical scenes.)

9. In this store, the molasses faucet made its bow to the world. Photo probably was made by Daniel Drawbaugh in the 1880's. The broken dry-plate negative was found among his effects almost a half-century ago.

Daniel's position with the pump company was that of master mechanic, designer of new inventions, and workman (when needed) at three dollars a day. The scope of operations expanded to include the manufacture of his stave cutter, stave jointer, mill tram, and an occasional steam injector. In 1870, the company incorporated under the name, The Drawbaugh Manufacturing Company.

Daniel Fettrow operated the village blacksmith shop in a small building adjoining his rented semidetached home. Landlord Elisha Hauck decided he wanted to sell—for $2,310. In March, 1868, Daniel Drawbaugh made his first real estate investment. On April 1, he took possession of the east side of the property, with blacksmith Fettrow moving his family into the other half of the house. Fettrow agreed to pay $110 a year as rent to the new landlord.

From the sale of his faucet patents and the house rents he collected, Daniel received some $7,000 during the years 1867 to 1870. He also received additional money from his jobbing work.

With his financial burdens somewhat lightened, he could, with an easier mind, experiment with his jelly glass and mustard can in his private room, and hope to solve the riddle of the transmission of speech over a wire—a "talking machine."

Just how successful he was in those early days cannot be stated positively in the light of much conflicting evidence which would come later—and, we do know that others were searching at the same time for the same answer.

Six years before the organization of the pump company, an incident of some scientific importance occurred in Europe. On October 22, 1861, a German, Philip Reis, instructor of natural science at Garnier's Institute at Friedrichsdorf, addressed his assembled brother-members of the Frankfort Physical Society. His subject was "On Telephony by the Galvanic Current." Said Reis: "I have now succeeded in constructing an apparatus by means of which I am in a position to reproduce the tones of diverse instruments and even to a certain degree the human voice."

His instruments were exhibited and demonstrated, and a report of the proceedings was printed in the society's journal

10. The Eberly's Mills house and shop of John Drawbaugh, purchased by him from David Emmet, June 10, 1834. Here Daniel lived with his parents until 1856. On May 13, 1865, John and Leah Drawbaugh, then in Grundy County, Illinois, transferred title to Elisha T. Hauck. The acknowledgment, taken there, bears the wife's (x) mark. Daniel acquired it March 13, 1868, and lived there with his family until 1876.

11. The "old curiosity shop" along the Yellow Breeches Creek housed the Drawbaugh operations from 1868 to 1911. Here his telephones were developed.

Drawbaugh's Early Patents

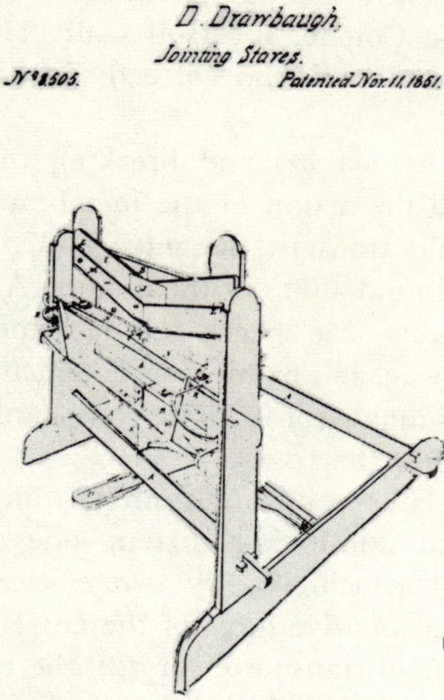

12. His first stave jointing machine "for casks of 15 different bilge." One of the four patents in that category issued to inventors in 1851.

13. The miraculous measuring faucet – molasses and financial support in more than just a trickle.

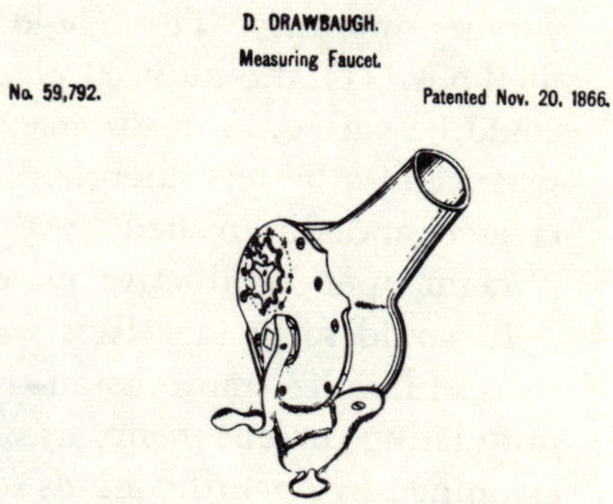

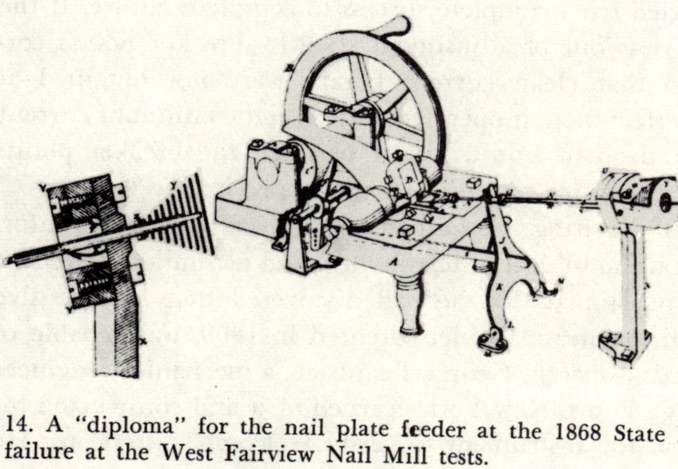

14. A "diploma" for the nail plate feeder at the 1868 State Fair—but failure at the West Fairview Nail Mill tests.

for 1860-61. Reis died in 1874. In 1888, at a rehearing of the 1885 Patent Office interference action between McDonough and Gray, and Bell and Edison, Colonel Ingersoll said: "He [Reis] had the art of telephony clearly and correctly in his mind." But did he?

Reis's transmitter operated by making and breaking the electrical circuit in response to the action of the membrane diaphragm. Such a device could transmit the pitch and, to some measure, the intensity of sound—but not the quality. All three factors must be present for clear speech reproduction. Reis successfully transmitted music, and by his own statement, "even to a certain degree the human voice." The consonants came through fairly well, but not the vowels.

The solution was discovered later to be a transmitter which could supply a continuous but undulating current—one of varying intensity. This could be achieved by two efficient methods: (1) the induced electromotive force of the current could be varied, as in the magneto transmitter or (2) the resistance of a battery energized circuit could be alternately increased and diminished by the action of a transmitter diaphragm upon conductive material.

It would seem that Reis was close to the achievement of successful voice transmission—but not close enough. After the undulating current theory was understood, his transmitter was examined by scientific minds to see just what it could do. Reports varied from complete success to complete failure. If the machine was out of adjustment slightly—breaker points corroded so that clean current breaks were not obtained—it talked better; then, it operated with a semi-continuous current effect. A drop of liquid placed between the breaker points would cause it to transmit articulate speech very well.

Scientific writings of the past years reveal interesting information on the subject of telephones—and facsimiles. Alfred G. Holcomb claimed that he had discovered that his sensitive polarized magnet machine, patented in 1860, was capable of transmitting speech. George Beardslee, a mechanical engineer of College Point, New York, learned of it and constructed his version of the instrument to verify Holcomb's assertions. He stated that his instruments demonstrated that the human voice could be conveyed by the method pointed out by Holcomb but viewed the machines as toys and took no further interest. Years later, it was shown that they were good receivers but, of course, at the time of their inception lacked transmitters capable of directly regulating a battery current.

In the Molecular Telephone Company case, the defense argued that the Beardslee instruments defeated the novelty of the Bell patent but Justice Wallace held that the work of neither inventor had presented sufficient evidence to overthrow the presumption of priority and validity arising from the grant of the telephone patent to Bell. An electro-phonetic receiver was the subject of a patent granted to Royal E. House in 1868. Designed for use on telegraphic circuits, it could have functioned as a telephone receiver. In 1869, Dr. P. H. Van der Weyde produced instruments of the type devised by Reis. They were demonstrated in New York City, and *The Manufacturer and Builder* newspaper pictured them and described their operation in May, 1869.

Just how much knowledge, if any, Drawbaugh had of the work of these and other pioneers, Reis excepted, is not known with certainty. By his own statement he had but little book learning. His formal education was limited to three to five terms of common school and, as he put it, "When I went, they were very common." He possessed less than a dozen books on scientific subjects, an English encyclopedia of several volumes borrowed from a friend, and a book on telegraphy. Prior to the spring of 1860, he attended a course of Friday evening lectures on physical science conducted by Professor Samuel Heiges, in the one-room Cedar Run Schoolhouse. To him, Daniel confided his dream of "speaking through the telegraph wire by electricity."

The leading scientific journal of that day was the *Scientific American* which began publishing about 1845. This periodical reviewed patents issued, printed plans to show the working of various devices, supplied formulae, and maintained a question and answer department for mechanical problems. Daniel was a subscriber "at an early period," as he put it, but could not fix the years. He did recall receiving copies late in 1863. The issue of March 4, 1876, carried a descriptive article on the Reis

15. An early model of the Reis telephone. It could sing, but could it talk?

16. The Holcomb magneto telephone (?), 1860-61.

17. Coffee mill? No. It's the Beardslee telephone (?) of 1865.

18. The Royal E. House patent of 1868. It could have been a telephone.

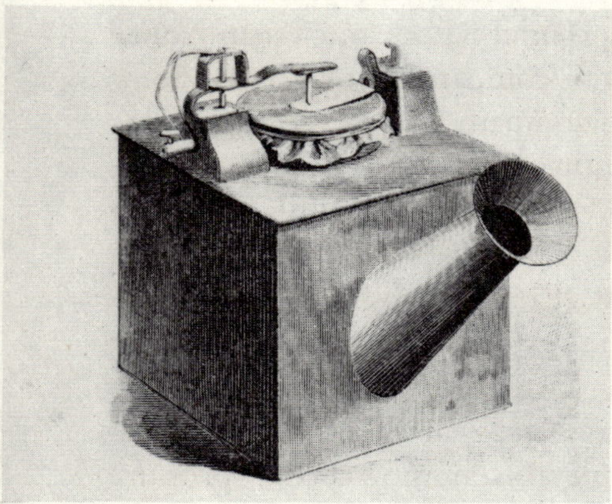

19. It looks like a teapot but it is the Van der Weyde transmitter of 1869—American version of the Reis instrument.

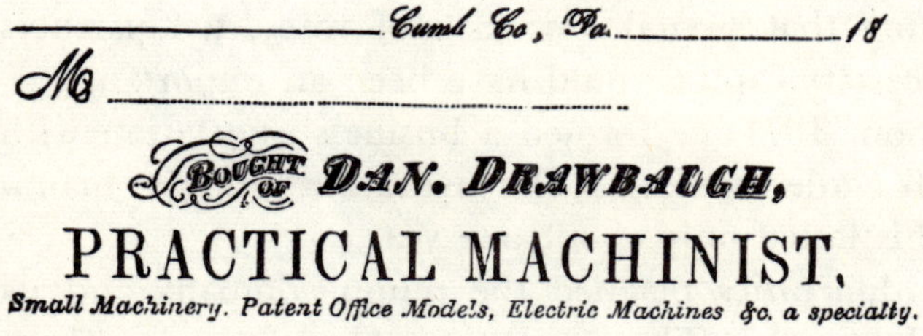

Cumb Co, Pa 18

M

BOUGHT OF DAN. DRAWBAUGH,

PRACTICAL MACHINIST,

Small Machinery. Patent Office Models, Electric Machines &c. a specialty.

20. Billheads were sometimes needed—customers occasionally paid.

telephone, and we do know that he had a copy and read it. The issue of September 9 of the same year gave a description of Bell's instruments with a somewhat inaccurate account of the theory of their operation. The following March 31 number ran an article with detailed sketches of the instruments covered by the second Bell patent, and almost any reader with electrical knowledge would have been able to construct a pair.

Daniel claimed that he began his experiments sometime prior to 1866 (his statement concerning daughter Emma placed the date as early as 1863) with a variable resistance transmitter composed of a dampened "low conductor" powdered material in the bottom of a tea cup—crude in form but correct in theory.

J. H. Teale, owner of a neighboring gristmill, patented a mill-bush, or bearing, in 1867. The following year, with William Sadler, Thomas Bergner, Daniel Rupp, and a man named Landis as partners, he formed a company to market it. Some $1,500 was paid to the pump company for special machinery which Daniel designed and built. This was set up and operated in the creek workshop, and the item was manufactured for about one year.

Between the years 1869 and 1875, the "wizard" was experimenting with various machines and devices. Among them were the steam injector, an automatic boiler feeder, paper bag folding machine, weather forecaster, carpet rag needle, machine for insulating wire, metal stamping press, pressure blower, two stamp cancellers, hydraulic ram, electro-magnetic clock, magneto telegraph key, automatic fire alarm, writing telegraph, magneto dial telegraph, a phonograph, and, of course, his beloved "talking machines."

Apparently none was developed fully since he received no patents during that period—but lack of interested partners with the necessary capital could have been an important reason, too. About 1871, he formed a business association with young Jacob Sadler to install the hydraulic rams he manufactured. This lasted only about one year.

Financial difficulties plagued the pump company and its incorporated successor, The Drawbaugh Manufacturing Company, from beginning to end. In March, 1869, $2,000 was borrowed from Henry Attick followed by loans from Messrs. Coble and Ditlow. At a meeting of the company, March 25, 1870, certain individual members agreed to personally repay the loans. This was not sufficient to free the company of debt, and an assessment of $30 was levied on each share of stock. A few months later, it was necessary to arrange a bank loan for $1,000 and give a sixty-day note to Brady, Mishy and Company for $475. In December of the same year, the stockholders were assessed an additional $10 a share to be paid by the twentieth of that month without fail. So went the company's progress—or, more correctly, the lack of it.

David Hauck must have appeared as an angel of light to the harassed management when he walked into the shop in June, 1873, with a view to purchasing the equipment and business. He represented Hauck Brothers and Company of nearby Mechanicsburg. Subsequent negotiations resulted in the sale for $7,000 cash. The purchaser took possession July 15. William R. Gorgas, the president of the liquidated corporation, later said his loss was $4,000.

The winter of 1875-76 saw an event which would shape the future of Daniel Drawbaugh—and, involve him in years of litigation. An obscure Boston speech instructor filed for letters patent on an "improvement in telegraphy." Tuesday, March 7, 1876 (a few days from his twenty-ninth birthday anniversary), patent No. 174,465 was issued to him. Alexander Graham Bell then was the owner of the "most valuable single patent ever issued."

As a mechanic and machinist, Drawbaugh ranked with the best, but always refused full-time employment with firms which offered excellent pay for his services. This was much to the despair of his wife. He would not go with them; he said he could earn more at job work, so they came to him and his shop always was tenanted by one or another manufacturer. The Haucks still were manufacturing faucets on the first floor when new faces appeared on the second.

Jacob Kline of nearby Camp Hill owned patents on a journal for sewing machines and agricultural implements, and on a wagon axle—but he wanted associates with capital. That want was filled December 23, 1874, with the formation of a copartnership which included Wilson Bear, M. M. Grove, Kline, and his son-in-law, Captain J. A. Moore. The enterprise, legally named, Bear, Moore, Grove and Kline, also was known as the M. M. Grove Company and the Harrisburg Axle Company. Moore, the moneyed man of the group, operated the Soldiers' Orphan School in what now is the 2100 block of Market Street, Camp Hill.

Satisfactory arrangements were made for part of Drawbaugh's shop; he was engaged as chief machinist; considerable new machinery was purchased and installed; and in the midst of the 1874-76 depression, the new enterprise was started.

Moore had sold his interest in the orphan school to devote full time to the manufacturing business. This was an ill-advised move. The history of each occupant of the creek workshop proved to be one of financial hardship—the axle company was no exception. On November 18, 1876, the partners formally transferred all property to Moore, the only solvent member, to liquidate. He did this in a thorough, efficient manner and bought back his school the following March to return to the lesser, more lucrative worries of managing, guiding, counseling, and caring for 234 orphans.

In his testimony, Danièl claimed he would have taken a patent on his telephone as early as 1870 or 1871 had he the means. As has been said already, we do know that for the years 1867-70 he received some $7,000. His magneto instruments "C" and "I" (see pages 60-61) were in existence then. He had worked "C" as a transmitter in conjunction with "I" as a receiver with fairly satisfactory results—according to his testimony.

A patent would have cost $60 or less—a caveat, just $15. But, nothing was done.

For the years 1871 through 1876, he experienced another lean period in his life with little money coming in. He did receive some $1,900 from the rents he collected, from his share of the proceeds of the faucet company liquidation, and from the wages he received from the axle company. He also received an unrecorded amount for his jobbing work done.

It wasn't much return for the effort expended and he was deeply in debt, too. But he had credit and friends and then,

21. Reading, writing, and arithmetic for the Drawbaughs—father and children—at the Cedar Hill or Gorgas School. Daughter Maude appears in this 1885 view.

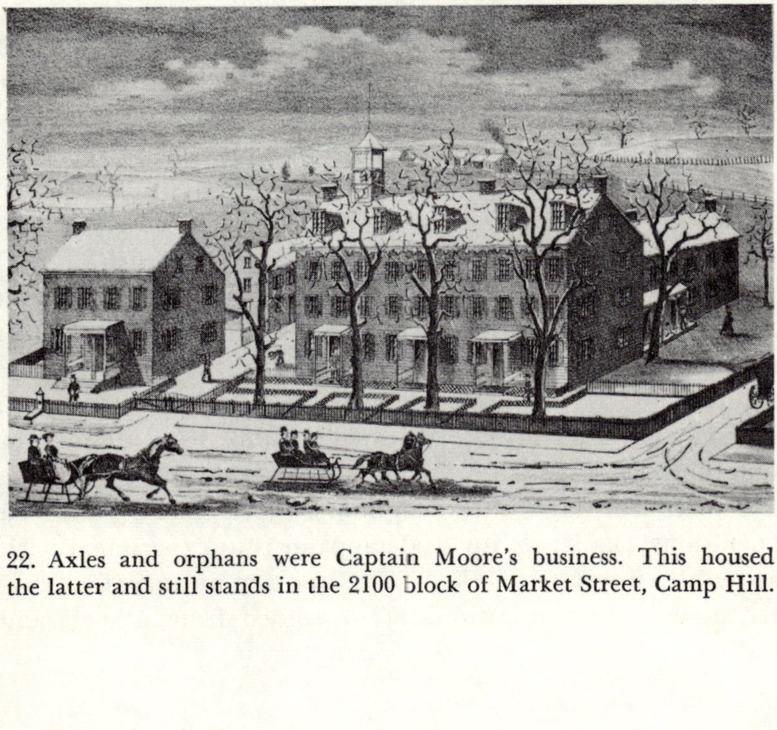

22. Axles and orphans were Captain Moore's business. This housed the latter and still stands in the 2100 block of Market Street, Camp Hill.

of all times, he should have used those resources. He didn't, however, but rather thought to dispose of his homestead to raise money.

But Elcetta stood firm. In their poverty, she regarded this one holding, encumbered as it was, as her only safeguard against the poorhouse. She would release her dower-right and permit a sale only if another house were purchased.

Over in Mechanicsburg, on East Main Street, Eli Wise had a nice, single, brick house which could be bought for $1,740—$1,000 down, and a series of three notes for the balance would do it. Low down-payments and F.H.A. thirty-year mortgages were still things to come.

During the previous eight years when blacksmith Fettrow had been the tenant of, and the next door neighbor to, the Drawbaughs, he had managed to feed his family, sometimes that of his landlord, and to hammer out a profit on his anvil.

In him, Daniel had a likely prospect and a deal in which Elcetta would concur. Negotiations resulted in a sale for $2,200. A title search revealed a lien of $1,000. Purchaser Fettrow assumed it, paid a cash balance of $1,200, and took possession April 1, 1876, when the Drawbaughs moved to their new home.

Living in Mechanicsburg proved to be fine—but awfully inconvenient for Daniel. The shop was still at Eberly's Mills, a good mile's walk from the Cumberland Valley Railroad Station at White Hill. Then, there was the bother of the daily train ride—and he couldn't get to the shop at odd hours as in the past. For a while, he lived with his brother in the village to be close to his experiments.

Finally, Elcetta gave up and the family moved back to Eberly's Mills in April, 1877, into the miller's house across the road from the old gristmill. A year later, they moved again, into the house behind the mill. Daniel was right back where he had started twenty-one years before—but with a family of six children by now.

Sometime in 1868, after the pump company had commenced operations, Daniel set up crude apparatus to experiment with an electro-magnetic clock. The temporary arrangement consisted of a wooden pendulum suspended from a shelf and fitted with a permanent magnet on the lower end. Beneath this on the floor was an electro-magnet. Copper and zinc plates were inserted in a box of moist earth to supply a feeble current to the electro-magnet. A "make-break" switch was fitted to the top of the pendulum to open and close this electrical circuit as the pendulum swung—and swing it did.

A complete clock then was built and enclosed in a rough pine case. Outside the shop, copper and zinc plates were buried in the earth to provide an unending supply of current. This model was set running about 1869-70 and operated intermittently until 1876.

That year, improvements were made to the mechanism; a walnut case was provided and the clock taken to Rufus Shapley's jewelry store in Mechanicsburg for exhibition—and to help him make up his mind as to the investment of $2,000 he then was considering. Again Daniel's luck held true to form. Shapley invested his money in the construction of his own version of an earth-operated timepiece.

Undaunted, the "wizard" proposed to produce one better still and experimented with a master clock to drive slave unit timekeepers, visioning the clocks of a city controlled by one timepiece. He thought to eliminate the hourly chimes, and by constructing a device to be used with either the phonograph or the telephone, cause his clocks to announce the hours vocally!

A fourth clock was made with brass works, compensating pendulum, and other refinements, and assembled in an ornate case for exhibition purposes. And exhibited it was, in a Harrisburg storeroom at 332 Market Street during May, 1878. Glowing newspaper accounts heralded its appearance, and it was suggested that one be made and placed in the dome of the state capitol.

All this was good publicity for the newly formed "clock company" just setting up in the 'Old Curiosity Shop," as the building along the creek came to be known.

Just the month before, the Grissingers, Theodore and Jacob, along with their brother-in-law, J. E. Shettel, had pooled their resources to buy three-quarters interest and to provide cash to manufacture and market the timepiece. It wasn't successful;

two were made and operations suspended. One was placed in the home of dentist, John W. Moffitt, on Market Street, Harrisburg, who wrote them to the effect that if they "could not make it run better, they should take it away."

But, Edgar Chellis, operator of a 99-cent variety store on North Third Street (in the Grand Opera House building) and Daniel's partner in faucets and water motors, kept the faith. He bought the first one and remained impressed enough to buy into the practically defunct clock company, in late 1879 and early 1880, by acquiring the interests of Daniel's partners. The inventor had obtained a patent on an improved earth battery for the clocks in 1879, and continued his experiments to better the accuracy of the timepiece. With every thunder shower and the accompanying magnetic disturbance to the earth, the clock ran wild, but—not everyone found time heavy on his hands.

Back in 1869 when the pump company was manufacturing faucets in the creek workshop, inventor Drawbaugh was studying ways to improve a pressure blower. He soon saw the possibilities of using it to draw and measure viscous fluids. It was similar to the faucet then in production but with a difference insofar as the attachment of the cut-off blade to the rotating piston. By incorporating this change, an improved faucet could be made at less cost. The company considered its manufacture—but wanted the rights without cost. Said Daniel: "I didn't feel to give my ideas so cheap as that," and put aside the device to await the coming of partners who would pay.

In December, 1878, dentist Moffitt, who didn't like magnetic clocks, talked with his friend, storeowner Chellis, who did. Both had money to invest in inventions, so they compromised —ruling out telephones and buying the faucet. The following January 15 found Daniel's patent application in Washington.

Over at Mechanicsburg, Hauck Brothers had been manufacturing the faucet purchased from the Drawbaugh company in 1873. David Hauck, part-owner and the machinist-inventor of the Hauck firm, had worked out improvements to this faucet. On January 17, he filed for his letters patent—two days after Drawbaugh.

The similarity of claims created an interference action.

Testimony was taken by both sides and the case reviewed in the Patent Office in early 1879. During this time, Moffitt, the partner of little faith, expressed his lack of confidence in Daniel. He sold his interest to Chellis. He should have waited, since the Drawbaugh claim was sustained, and on July 6, 1880, Daniel received his patent.

The same year, he received his papers on an improved water-powered motor. One of these later was installed to operate the pipe organ at St. Patrick's Pro-cathedral on State Street, Harrisburg.

Under the terms of the new Chellis-Drawbaugh agreement, the inventor was to devote his whole time and attention to the manufacture of the new faucet and water motors which also were to be made.

But, he didn't do it.

His mind was on his telephones. Chellis complained to Mrs. Drawbaugh, who already was well aware of the plaguing (to others) situation. Chellis reported: "Said she: 'Mr. Chellis, Dan works at the telephone as soon as you go away, and most every night does not get home 'til twelve or one o'clock.'. . . and, she said that she had been talking to him about it, and wanted him to lay aside the telephone and work on the patterns, and give up the telephone. He [Daniel] said he 'would go to the poorhouse before he would give up working on it.' "

Chellis, who also had tried to persuade him, but with no luck, said, ". . . I advised him to drop it—the telephone, as he could not antedate Bell." But, to that, the obdurate Daniel had one reply: "I don't know about that; I have been working at it a good while."

When a visitor happened in the shop one day and remarked that he remembered talking through it in 1870, that was enough for Chellis. He was ready to listen to reason.

Crowds packed the Centennial Exhibition at Philadelphia during the summer of 1876. The scientific exhibits especially were popular—but not Professor Bell's telephone. For some six weeks, it had been on display in an out-of-the-way corner at the east end of the main building, and attracted little attention. On the hot, Sunday afternoon of June 25, a group of scientific men and exhibition judges inspected Elisha Gray's harmonic telegraph exhibit and paused at Bell's display. Then came the inventor's opportunity on the heels of royalty. One member of the group, Dom Pedro, emperor of Brazil, strolled up to the table. He had met Bell once on a visit to Boston University, and remembered. His greeting was short but cordial. Invited to hear the new marvel, he placed a receiver to his ear, then dumbstruck cried, "My God, it talks!" At least, that's one version of the happening.

Aged scientist, Professor Joseph Henry took his turn listening as did Sir William Thomson, probably the foremost electrical authority of the day, while fellow-centennial judge, J. C. Watson, read excerpts from a newspaper. From that time on, every scientist visiting the exhibition made the speaking telephone a "must."

Daniel Drawbaugh had proposed to exhibit at the Centennial and by his later claims had made two magneto speaking instruments, "D" and "E" (see page 62) in February, 1875. They were good instruments, too, but rather feeble, and certainly the claimed completion date was sufficiently in advance of the opening of the fair to prepare. But what did he propose to exhibit? Holsinger, the Eberly's Mills printer, offered an answer to the question in his article on November 16, 1875, in the Carlisle (Pa.) *Mirror:* ". . . electric clock without a battery is being gotten up in our town by Daniel Drawbaugh to be exhibited at the Centennial next fourth . . ."

Why would he decide to exhibit a clock rather than a telephone? Most folks knew what time it was but who, then, had talked over a telephone? Furthermore, by later statements of the inventor and some of his witnesses, his telephone was fairly well developed prior to the Centennial, but the clock only in the experimental stages. He went to the exhibition on October 17, 1876, and stayed several days—but with neither clock nor telephone. Wealthy George Leonard, the village mill owner, accompanied him. Together, they saw everything there was to see, including the telephones on display. Drawbaugh said nothing of his telephone to his friend then, and later could not remember whether he had seen Gray's harmonic telegraph or Bell's telephone.

The inventing fraternity, fired with the news of Bell's suc-

23. That controversial clock—one of the original half-dozen known to have been built by Drawbaugh. It had the benefit of solar transit observations for regulation. Zinc and copper plates packed in coke and buried in the earth were required for its operation. The 25-pound compensating pendulum was controlled by magnetism—not by "maggots" as one countryman insisted. Long known of before our inventor constructed his first; always discarded because of unreliability; revived periodically as an object of curiosity—that controversial clock.

24. The reconstructed miller's house still stands across the road from the old gristmill site. Here Drawbaugh lived with his family for one year (April, 1877-1878). The testimony indicates the possibility that this was his first house to be connected with the creek workshop by telephone line (December, 1877).

cess, turned its attention to speaking devices. In 1876, there was one patent for an electric speaking telephone and Professor Bell had it. In January, 1877, he took out another on an improved telephone, and later that year inventor McTighe secured two patents relating to telephones. During 1878, there were 33 patents granted; in 1879, 64 granted; in 1880, more than 100; and in the first half of 1881, 94, and all on, or relating to, speaking telephones. Inventors everywhere bought wire, batteries, wound magnets, made diaphragms, shouted into horns—and hoped that something would come out the other end. Even the lowly farm hand—who had perhaps wrapped the handle of a pitchfork with wire to repair it—while he called in the cows, wondered if he didn't thereby make a speaking telephone, with which through some fluke he might antedate Bell.

After issuance of the second Bell patent in 1877, the development of the business began in earnest under the guidance of partner, Gardiner G. Hubbard. Bell assigned his interest to Hubbard as trustee in July, and then married his daughter.

The next year, Hubbard assigned the rights to the newly formed Bell Telephone Company, a Massachusetts corporation. When the American Bell Telephone Company was organized in 1880, the rights were vested in it to carry on the "manufacturing, owning, selling, using and licensing others to use, electrical speaking telephones and other apparatus and appliances pertaining to the transmission of intelligence by electricity, and for that purpose constructing and maintaining by itself and its licensees public and private lines and district exchanges."

The first lease for a paying telephone (a few had been put out on a free trial basis) was dated June 8, 1877. By January, 1878, there were 10,000 instruments in commercial use and the first telephone exchange as such was established that year. In late 1880, exchanges existed in more than 275 cities and towns, and in every state of the Union.

Most telephone conversations were more ordeal than pleasure. Circuits were aerial, supported on poles and mingled with, or in close proximity to, telegraph and arc lamp wires. The first lines had been constructed with a single wire; the earth provided the return circuit. Conversations had a background of other voices, clicks, and bubbling. If Professor Gray's harmonic telegraph was working on nearby Western Union wires, toots, wails, and catcalls were thrown in for good measure. A Philadelphia electrician, David Brooks, came up with a simple remedy for most of those inductive disturbances—forget the earth and use a second wire instead for the return circuit.

Those early months were anything but pleasant or profitable for the Bell promoters. There was little money to manufacture instruments, string wire, and provide service—to the few who would take it. Capitalists very definitely weren't interested in scientific toys—or in gambling with a Bell franchise. The man in the street saw the speaking telephone, talked through it, and decided he didn't need one in his home. Business firms held the same opinion. It might do, however, for lovers and diplomats, as one newspaper had put it.

Discouraged, the patent rights owners thought to sell out to Western Union, but its president, William Orton, wasn't interested—in buying.

However, Frank L. Pope, special advisory electrical expert to Western Union and its subsidiary Gold and Stock Telegraph Company, strongly advised that the management investigate telephone possibilities. From Europe came a copy of every writing of the past fifty years on electricity, telegraphy, and related subjects. A skilled scientific translator was employed to help Pope familiarize himself with the state of the art. Then, Western Union acted.

Thomas A. Edison, with Orton behind him, applied in July, 1877, for a patent on a carbon telephone transmitter far superior to anything then in use. Professor Amos E. Dolbear, physics instructor at Tuft's College, filed for improvements to magneto telephones. Claimed were iron diaphragm and permanent magnet features. Elisha Gray held some patents on inventions which he claimed were utilized in telephones. He filed three additional applications covering a liquid transmitter, magneto receiver, and other improvements. Late in 1877, the American Speaking Telephone Company was brought into being by Western Union money to provide telephone service, using the instruments and ideas of these men.

The Bell interests then had a decision to make—sit tight and run the risk of being frozen out or lock horns with this stepchild of the mighty colossus of the communications business and fight it out. They chose the latter course and brought suit in federal court for infringement of patent. Before a decision was rendered, Western Union decided to negotiate. An agreement, reached in November, 1879, stipulated principally that Western Union would retire from telephones, and that the Bell interests would stick to them and forget any ideas about entering the telegraph business. Western Union was compensated for the value of certain subsidiary patent rights, and the American Speaking Telephone Company, and its paying customers, passed into Bell control.

There were some 140,000 electric speaking telephones legally in use by October, 1880. Legally at that time meant under license from the Bell company on a royalty basis. When infringers popped up, the Bell lawyers cracked down. A. D. Hard of Boston, in 1878, made and sold completed telephones, and parts for later assembling. An injunction was obtained to stop this. He was ordered to, and did, pay damages. That year, Jerome Redding, J. H. Bunnell, and Jones Brothers had similar experiences. Then, there were Pearce and Jones, the Eaton Telephone Company, and Lazelle with his "Little Gem" instruments.

The nuisance value of these small-time infringers was promptly cancelled. But with the appearance of the People's Telephone Company came a really worthwhile adversary. Of the six hundred suits the Bell interests would fight in a quarter century, none would compare with the People's Company litigation for length and bitterness, except perhaps, the action of the United States government to revoke the Bell patents.

In his home on West Fourth Street, New York City, inventor Frank A. Klemm devised an improved battery transmitter and applied for a patent on it. On May 4, 1880, he received it. Another New York State inventor, Abner G. Tisdel, directed his efforts to ways and means of improving magneto instruments. This patent was issued May 18, 1880.

Tisdel assigned his rights to Klemm, Ernest Marx of New York City, Moritz Loth of Cincinnati, Ohio, and Simon Wolf of Washington, D. C. By Agreement of March 6, 1880, the assignees were required to hold these interests until the formation of a corporate body. Then they would transfer to that body.

The plan was to organize under the name of Klemm Telephone Company, construct lines and exchanges, and provide telephone service. Others, too, had tried the same thing, and it has been said that, in the three years following the Western Union case, one hundred and twenty-five companies were started in open defiance of the Bell patent. But, the Klemm group was determined to test the validity and scope of the Bell patent—if necessary. Counsel advised that probably the only successful defense would be that of prior invention. Some inventor with prospects of establishing that was needed. Someone who could antedate Bell.

iii

Chellis, Drawbaugh's investor-partner, had consulted with Harrisburg attorney Michael W. Jacobs, who had represented him and Daniel in the Hauck faucet interference case. Former Judge Lysander Hill, an eminent barrister of Washington, was called to Harrisburg for a consultation. A visit was made to the creek workshop probably in the late summer of 1879. Concerning this, Chellis said: "I took Mr. Hill over to visit Mr. Drawbaugh, and Mr. Drawbaugh showed him the telephones, and explained them to him, but Mr. Hill said it would cost $100,000 to fight the Bell Company; he considered Mr. Drawbaugh the original inventor, but advised us not to go into it unless we were prepared to raise $100,000 to fight the existing companies."

He might just as well have said a hundred million. Counsellor Jacobs had a small law office on North Third Street and a small income to go with it. Chellis could have run a special "49 cents marked down from 99 cents" sale at the variety store and raised a few hundred dollars. The third prospective client had trouble paying bills, let alone raising money. Henry Springer, the village shoemaker, who kept the Drawbaughs shod, had kept a running account, too. Eighty cents at a time was more than the inventor's budget could stand for shoeing.

To help him with his experiments, Simon C. Wilson, superintendent of telegraph repair for the Northern Central Railroad Company, supplied him with scrap electrical material, parts of batteries, coils, and magnets which the railroad gave away at its Harrisburg shop.

Said schoolmaster Daniel Hammacher: "During the spring term of 1879 and the spring term of 1880, I taught school at the Cedar Hill or Gorgas' School House. Mr. Drawbaugh sent his children to me to school and also, I purchased some books for those children, and it was a great deal of trouble to collect the money. I did not get it until the beginning of September, of the fall of 1880; I gave it to a constable . . . A. B. Shank . . . to collect . . . he collected the money; amount of all, books and tuition, $13 and odd cents."

Trite as it may seem, where there's a will there's a way.

Chellis, Jacobs, and Hill talked over the matter and decided to become partners with Daniel without the outlay of cash, so it was said. Under an agreement, dated May 6, 1880, each received one-quarter interest in the Drawbaugh telephone inventions (or "pretensions" as they were later termed by the opposition), and a hint was dropped in the right place.

On May 24, Simon Wolf arrived in Harrisburg and registered at the Jones' house. He was in town on a buying trip. Intangibles were his market—telephonic intangibles. Perhaps he could acquire all right, title, and interest in Daniel's telephones.

It would be interesting to know what was said by Messrs. Wolf and Chellis at that Monday meeting. In his testimony before Examiner Ott, July 2, 1881, Chellis tells us this:

(Q) When Mr. Wolf first came to you to negotiate for the purchase of the Drawbaugh invention, did he, or not, tell you that he represented a party who had already purchased the telephone inventions of Klemm and Tisdel, and were purporting to enter upon their manufacture and sale, and that they expected trouble with the Bell Telephone Company, who claimed to monopolize electric telephones of all kinds, and that they had already fully determined to test the validity and scope of the Bell patents, and the question whether the Klemm and Tisdel telephones were infringe-

ments thereof; what, if any, statements did he make on these several points?
(A) He did state something to that effect; it was in Mr. Jacobs' room at the hotel; I can't remember the exact words he said, it was to the effect that they wanted to use it in evidence against the Bell company; I mean they wished to use the Drawbaugh invention in evidence against the Bell company.
(Q) Did he, or not, state that he had been advised that a patent might yet be granted upon the Drawbaugh invention, and that, unless they purchased the invention, the Bell company might themselves purchase it, if they found their suit likely to be defeated by it, and might then get a patent on it themselves, which they could use against Mr. Wolf and his associates?
(A) Something to that effect, but I cannot remember the exact words.

Bargaining didn't prove too difficult, and by June 22, the Drawbaugh and Klemm interests had signed a conditional sales contract.

The grandest day of Daniel Drawbaugh's life came July 21, 1880, at Washington. All the instruments that had been brought along were photographed. Daniel filed for letters patent on his telephones and then sat down and signed away his rights to them for $20,000 and a very large, but never disclosed, amount of stock. The assignees of course were Klemm, Marx, Wolf, and Loth.

It is not clear just which instruments were included in the first patent application which was assigned serial number 14,040 and dated July 26. A second application, serial number 19,567, was filed October 26. The best information the writer has leads to the conclusion that the "H," "J," "F," "B," "C," "I," and "A" machines were included—probably in the two applications. At this date, it is impossible to obtain information concerning the applications. In keeping with the provisions of the federal records control schedule's ruling that all abandoned Patent Office applications be disposed of twenty years from the date of filing, they have been destroyed.

Financial circles buzzed with the news. Said *The Cincinnati Commercial* in its July 22 issue, under a July 21, Washington dateline:

An application for a patent was filed today that, in consequence of its vastness of interest, as well as wealth of prospect, renders it a subject of national interest . . . The company is composed of leading business men from all parts of the country . . . The cash capital of the company is $5,000,000, with headquarters in New York and in about sixty days they will open up the telephone, which will certainly result in the driving out of all telephones in the market, save the ones they hold, or else compelling the Gray, Bell and Edison lines to pay the new company a munificent royalty.

After this optimistic recital, *The New York Journal of Commerce* on August 23, 1880, reported:

A company has recently been formed in this city with a capital of $5,000,000 for the purpose of manufacturing telephones. The company is to be known as the People's Telephone Company, and a number of leading capitalists in this city and Cincinnati are interested in it. The telephones are to be manufactured under the patents of Frank A. Klemm and Abner G. Tisdel, and the application for patents of David [sic] Drawbaugh of Eberly's Mills, Cumberland County, Pa. filed July 21, 1880. It is claimed by those interested in the new enterprise that Drawbaugh is really the inventor of the telephone, and had completed one years before Professor Bell or any one else had manufactured one. He was, however in very humble circumstances, and his neighbors who knew of his experiments looked upon him as a harmless lunatic. He continued improving his original telephone, and it is claimed that the one which the new company proposes to furnish is superior to any now in use. The company has fitted up a factory in Brooklyn, and in three months will be prepared to supply 1,000 of the new telephones . . .

Daniel's assignation of his rights was the on-stage cue for the proposed corporate body waiting in the wings. The Klemm name was dropped and Joseph Loth, Julius Bien, Richard McCormick, Marcus Marx, and Auguste H. Girard filed their articles of association in the office of the clerk of the city and county of New York, August 30, 1880, under the name of People's Telephone Company. A few days later, on September 4, assignees Marx, Loth, Wolf, and Klemm transferred their rights to this new company.

Then the People's Telephone Company was a reality with a term of existence of fifty years and rights of "the manufactur-

THE AMERICAN BELL TELEPHONE CO.

95 MILK ST., BOSTON, MASS.

This Company owns the Letters Patent granted to Alexander Graham Bell, March 7th, 1876, No. 174,465, and January 30th, 1877, No. 186,787.

The transmission of Speech by all known forms of Electric Speaking Telephones infringes the right secured to this Company by the above patents, and renders each individual user of telephones not furnished by it or its licensees responsible for such unlawful use, and all the consequences thereof, and liable to suit therefor.

25. Advertisement of The American Bell Telephone Co.

26. But, Western Union, with its telephone instruments of Phelps and Edison and the rights of Gray and Dolbear, was hard to convince. This early-day exchange of the Merchants' Telephone Company at 198 Broadway operated under a franchise from Western Union's subsidiary company until late 1879, when it passed into Bell control.

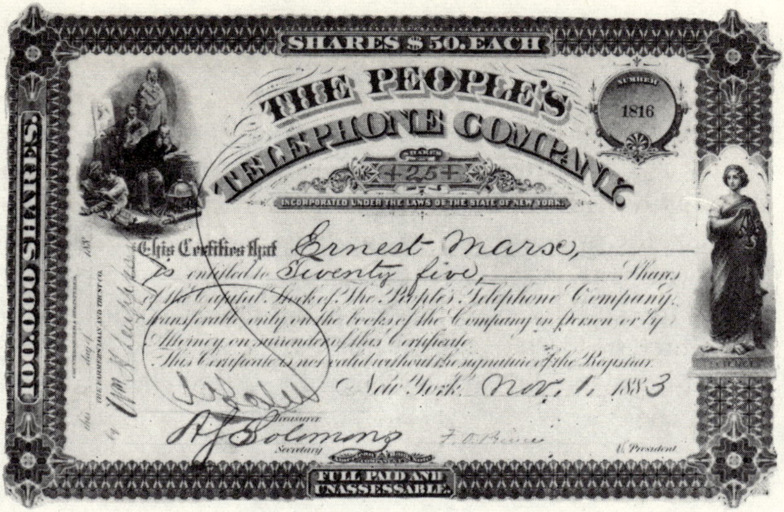

27. While awaiting the court decision, shares and franchises were disposed of.

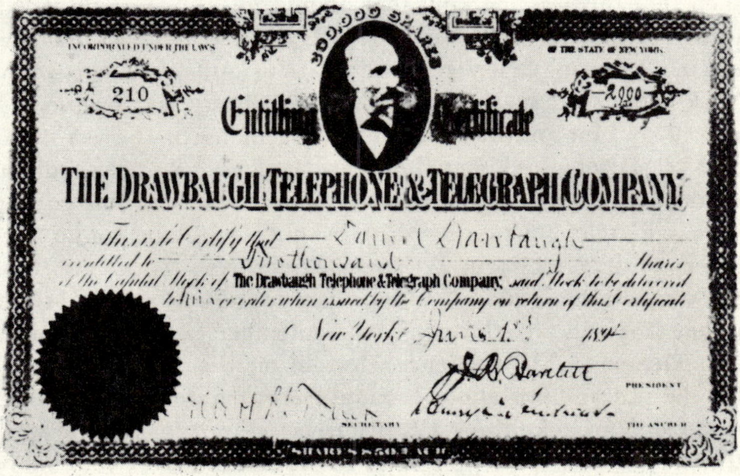

28. And the Drawbaugh descendants still hold their certificates.

ing, constructing, owning, furnishing, maintaining, letting and selling telephones, and the apparatus used therewith, under the inventions and patents of Abner G. Tisdel, Frank A. Klemm, Daniel Drawbaugh, and other inventions and patents which may hereafter be assigned to said company, and the doing all things to said business or necessarily connected therewith . . . the principal place of business and office of said company will be the City of New York . . ."

Edward L. Bradley, a modest, unassuming gentleman, became interested in telephones—interested enough to go around, look at them, and buy a few. Concerning his purchases, he had this to say:

On or about June 11, 1880, I purchased of Frank A. Klemm, personally, at his residence, 234 West 4th Street, New York, a pair of magneto telephones . . . On that occasion Mr. Klemm told me that he and those associated with him, intended to make, and put out for use by the public generally, electric speaking telephones, and to establish by themselves and those whom they would supply with telephones, telephonic exchanges; . . . that he was endeavoring to perfect a battery-transmitting telephone, to be also sold and put out the same way. . . . On July 31, 1880, I also purchased . . . of Klemm . . . [a] Klemm battery telephone . . . . He told me then that this was the kind that he and his associates intended to put out; that he expected to supply such telephones for use as electric speaking telephones, in Harlem and elsewhere in New York City. That such telephones, both battery and magneto, furnished by him and his associates, were in use in New York City and at Harlem, and that they were arranging to sell them to be used in other parts of the United States . . . That a line equipped with said telephones, furnished by him, could be found at the place of Mr. Groesman, 211 Canal Street, New York City. I went to . . . Groesman's place . . . and there found one station of a telephone line and talked over it. On September 11, 1880, I again saw Mr. Klemm at [his] residence. He told me that a company got up for the purpose was about to manufacture and sell electric speaking telephones like those I had purchased of him; that he was interested in said company; that Mr. Richard McCormick of New York, was its president, Mr. Loth of Cincinnati, was treasurer, and that Mr. Marx of New York, was its manager; that an executive office for it was being fitted up at 4 East 14th Street, New York;

that as soon as arrangements could be completed, they should endeavor to establish exchanges and sell telephones. I understood from him that they had applied for patents on some alleged inventions of Drawbaugh. The building [at] 4 East 14th Street . . . has at the street door a sign, "People's Telephone Company." Up one flight of stairs there is an office with the same name on the door. About the middle of September, 1880, I called at that office and found Mr. Marcus Marx there. In conversation he told me that the People's Telephone Company had begun to furnish telephones and were [sic] putting in a few, but refused to tell me where . . . I asked him to let me see one of the telephones they were putting in and he said he would do so later. I called on him again on September 25, 1880, at the same office. He declined to show me any telephones, or to tell me where he or his company had put them up, or to give me any information about the subject, stating as a reason that he knew I was employed by the Bell Telephone Company.

After examining Drawbaugh's history, the Bell Telephone Company went into the United States Circuit Court of the Southern District of New York with a bill of complaint. This was October 20, 1880, just three months after the grandest day in the life of Daniel Drawbaugh.

Appendix A. The early Drawbaugh telephone instruments

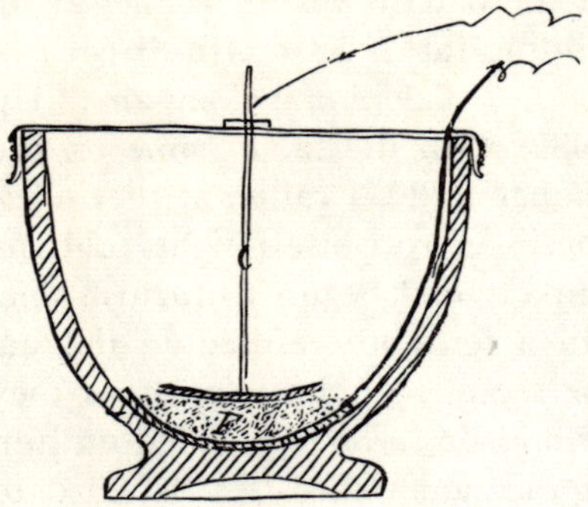

29*a*. Drawbaugh sketch 1—the teacup transmitter.

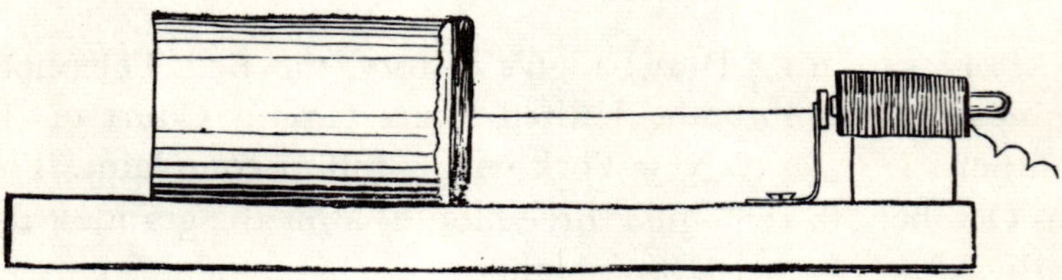

29*b*. Drawbaugh sketch 2—the first receiver "B"—claimed for late 1866.

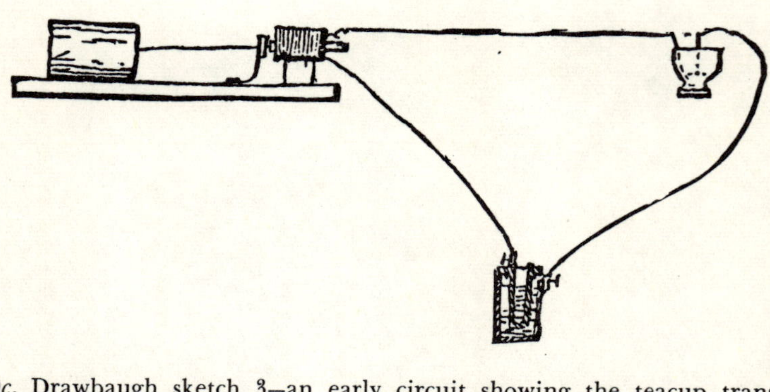

29*c*. Drawbaugh sketch 3—an early circuit showing the teacup transmitter, mustard can receiver "B," and the battery.

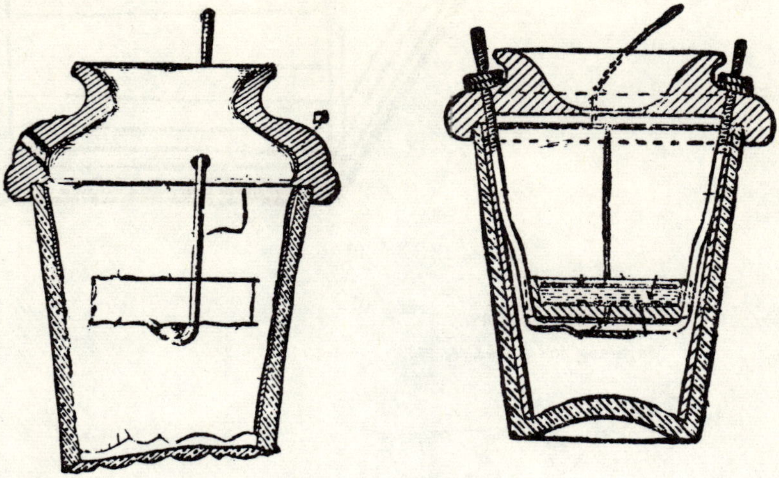

29*d*. Left, "F" original—the first jelly glass transmitter. Claimed for early 1867. 29*e*. Right, "F" improved. The stirrup rod and metal diaphragm had been added along with an unbroken jelly glass. Claimed for an unknown time soon after the first "F."

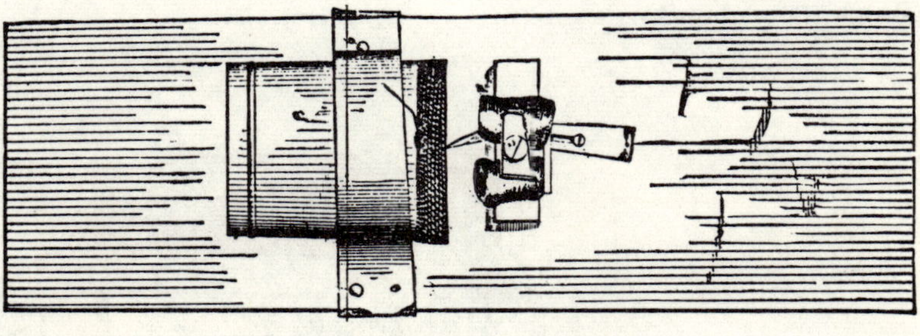

29*f*. "B" improved version—patterned after the first "B" instrument but with the magnets mounted close to the can, thus eliminating the string (compare with Drawbaugh sketch 2). Claimed for early 1867.

29*g*. Drawbaugh sketch 4. A paper ear cone to slip on the end of can "B."

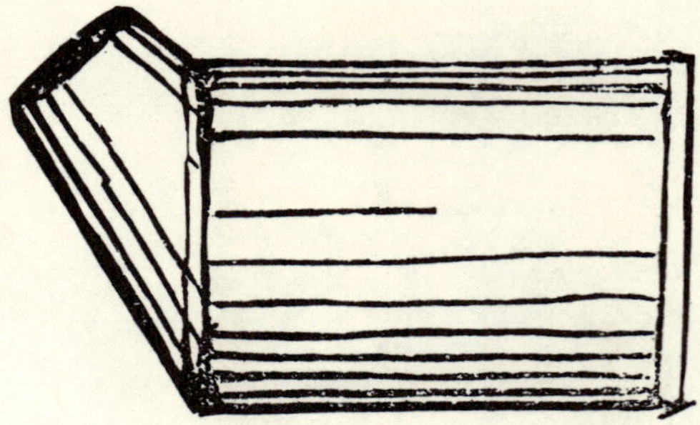

EXHIBIT LORY.

[*Respondents' printed record, page* 236.]

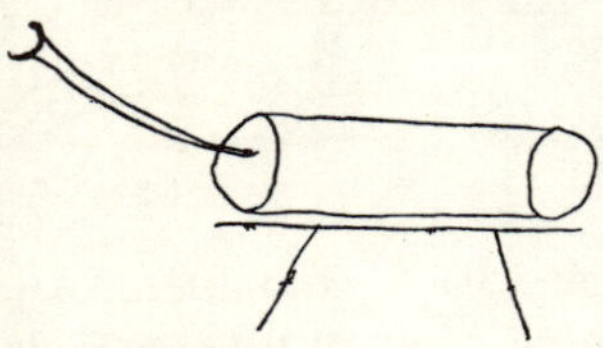

[*The above is a reduced photo-engraving of the sketch made by respondents' witness, W. M. Lory, May* 7, 1881, *in answer to direct question* 13. *The original sketch was made in pencil by the witness, and was inked in by Mr. E. H. Rydall, at the time, by consent of counsel. It is reduced to about one third of the size of the original.*

FRED. M. OTT,
Examiner.]

29*h*. The W. M. Lory exhibit. Drawn by the witness from memory of a sketch shown him by the inventor in the fall of 1864.

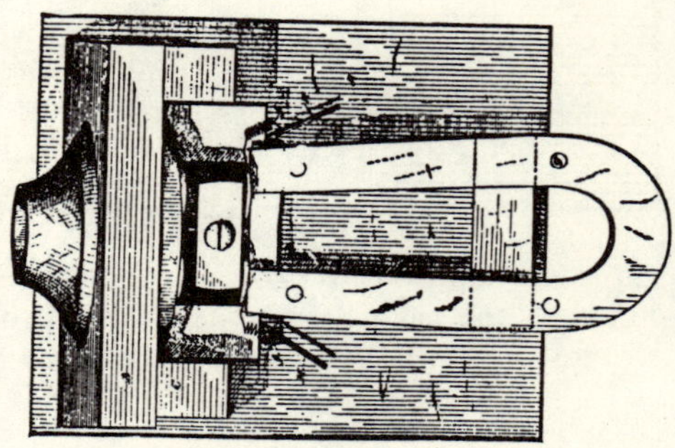

29*i*. "C reproduced"—a magneto transmitter-receiver. The original was first used as a receiver for the "F" transmitter. Claimed for 1869-70.

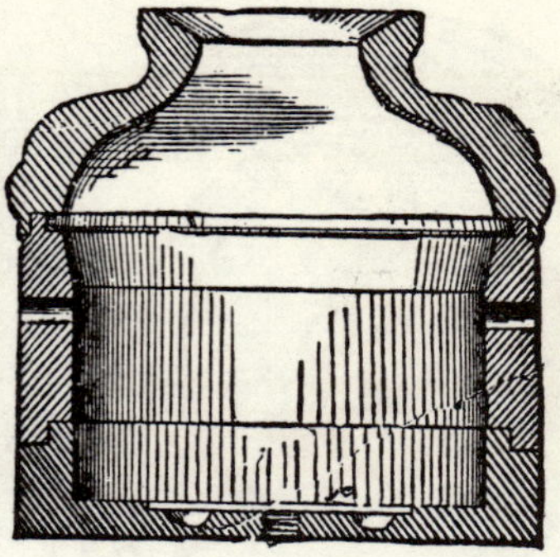

29*j*. Left, "I" as it came into the case—an empty box—would have been an excellent "string telephone" component.

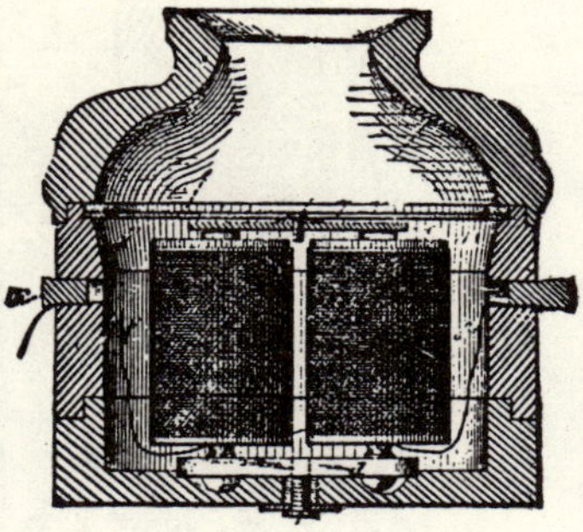

29*k*. Right, "I reproduced" by Drawbaugh as he said it was in 1870. Originally used as a receiver with "C" as a transmitter.

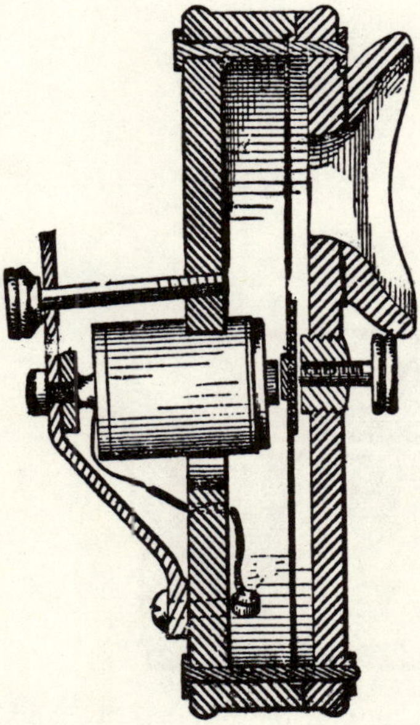

29*l*. "A"—a non-polarized receiver. Required a battery energized transmitter such as "F" for operation. Claimed date—fall of 1874. An earlier model with a centrally located ear cup was claimed for 1873.

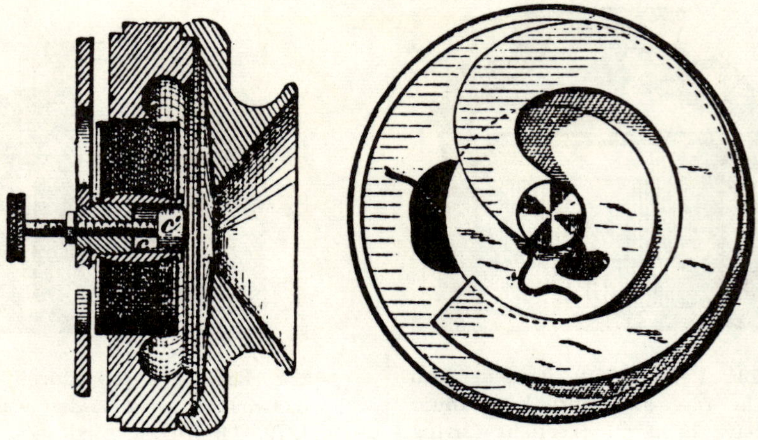

29*m*. "D" and "E"—similar magneto instruments for talking or listening. A pair would operate without a battery. Note the ear-shaped, permanent magnet typical of these particular instruments and shown in the right view. Claimed for January-February, 1875.

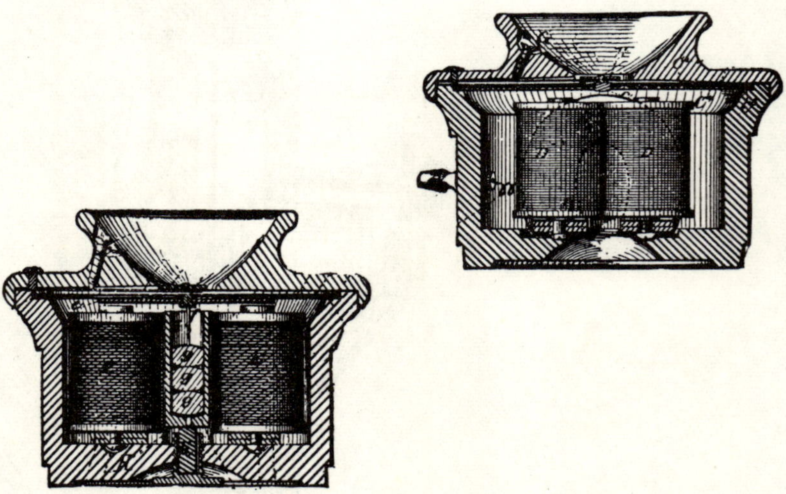

29*n*. Left, one of the original "L" and "M" similar combination, carbon microphone-magneto instruments for both talking and listening. Not practical. Right, view shows one of the instruments as reconstructed. Note that the carbon ball microphone section has been removed and the magnet coils placed together. The original instruments were claimed for early 1876—the rebuilt version soon after.

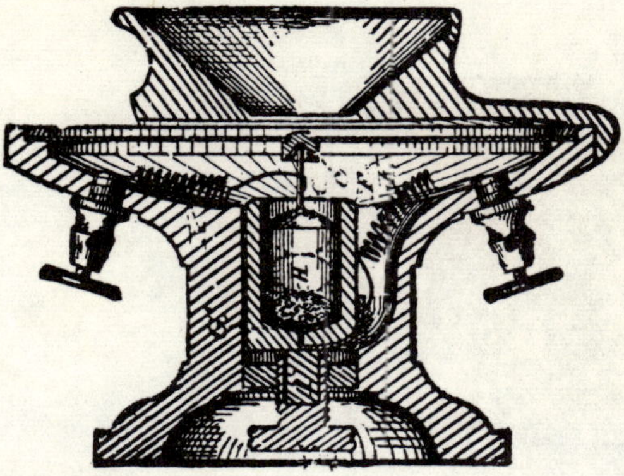

29*o*. "G" and "O"—similar hard carbone microphones, one of which is shown. No receiver included in the construction. They may be considered the first successful variable resistance transmitters made after the "F" jelly tumbler. Claimed for the summer of 1876.

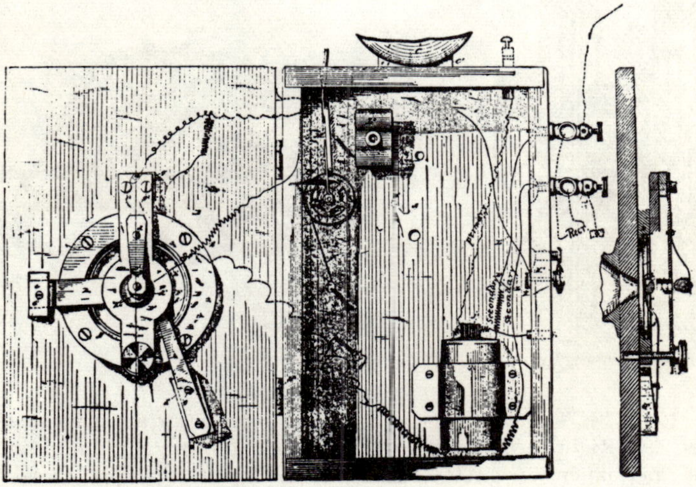

29*p*. "H"—a carbon microphone transmitter for wall mounting. Originally equipped with a call bell. An induction coil was mounted in the base of the box. Claimed date—before September,1876. Similar in features and appearance to the Bell-owned Blake transmitter.

29*q*. "J"—similar to "H" except that it had two sets of carbon points which were actuated by the diaphragm. Not practical, it had been laid aside. Made in 1878.

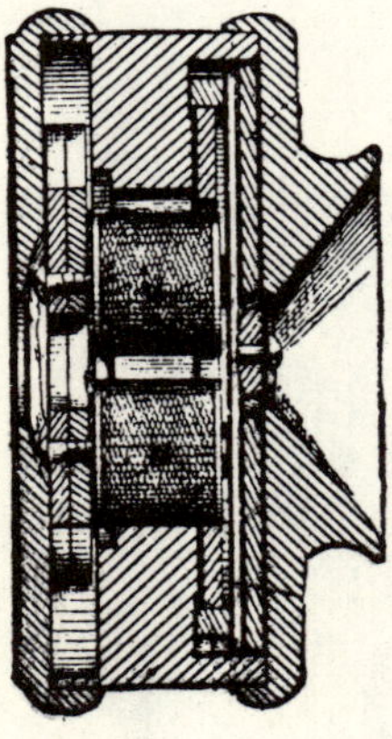

29*r*. Left, "N"—a double-pole magneto receiver with efficient short coils. 29*s*. Right, a photograph of the back showing the double horseshoe, permanent magnets. This machine featured a split metal diaphragm, the two halves of which are shown on the table. Claimed for the spring of 1879.

29*s*. The "Q" magneto machine (not pictured) was similar to the "N" shown above but employed a partially split diaphragm. Claimed for the spring of 1879.

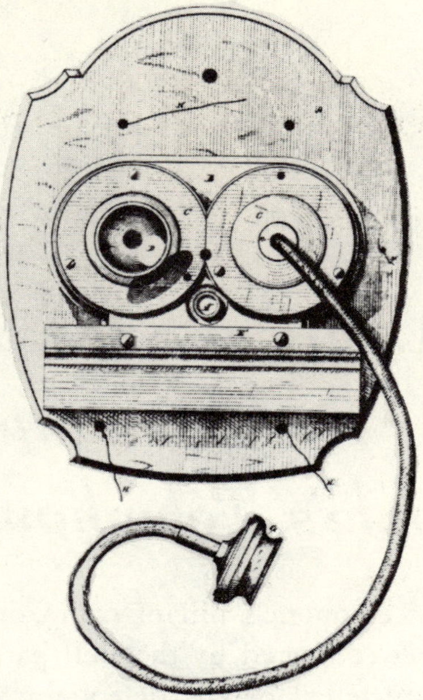

29*t*. "P"—a complete telephone station consisting of a carbon ball transmitter and magnet receiver. A rubber tube, encasing a spiral spring, extended from the receiver diaphragm to an ear piece equipped with a second diaphragm. Claimed for the spring of 1879, Daniel considered it poor and laid it aside.

29*u*. "K"—an instrument so impractical it was never completed. A clockwork movement was to revolve an inside cylinder by means of the long rod at the right. When the action was tested by hand, the resulting noise drowned the speech. Probably made in the summer of 1879—perhaps suggested by Edison's 1877 "Motophone."

II

American Bell Telephone Company vs. People's Telephone Company

Bell Telephone's bill of complaint recited the rights and privileges conferred by the Bell patents, declaring that these were threatened by the defendant who proposed to deprive without license or consent; that it was aware the complainant's patents would be infringed; that it was alleged and pretended that Daniel Drawbaugh was the original inventor of the electric speaking telephone; that the defendants associated themselves for the purpose of manufacturing, selling, and using telephones under the pretended inventions and improvements of Klemm, Tisdel, and Drawbaugh; that the pretended inventions of Drawbaugh had not been introduced into public use and that knowledge thereof had been withheld from the complainant except as disclosed by recent newspaper publications; that the defendant had been requested to acknowledge claimant's rights and to refrain from said acts and from disregarding and infringing the complainant's rights, and to account for and pay what was justly due by reason of the infringement, but that the defendant had so refused to do.

The plaintiff asked the court to make the defendant account for, and pay, the monies due, and to be restrained from further operations by writ of injunction.

On the motion for preliminary injunction, five affidavits in reply were filed by members of the defendant company. Statements of Richard McCormick, president, A. S. Solomon, secretary, Marcus Marx, member of the board of trustees, Simon Wolf, stockholder, and Lewis Abraham, counsel for the company, summed up to this: *The Cincinnati Commercial* and the *New York Journal of Commerce* articles were published without authority of the company and before its organization; the company never had made, sold, or used any telephones and had no intention of entering the business before its rights were determined in the Patent Office. No factory had been established in Brooklyn, or elsewhere, to manufacture instruments. Responsibility was disclaimed for certain statements of Marx, and for certain statements and acts of Klemm, and for ownership of the telephone line at 211 Canal Street as deposed by Bradley.

At the hearing, the People's Company in defense rested solely on the question of non-infringement, but the court adjudged the Klemm and Tisdel machines as infringing the Bell patents. On November 22, 1880, Justice Blatchford granted the injunction.

This opened years of legal processes, complete with volumes of testimony from scores of witnesses, tests of instruments, charges and countercharges, facts and mere opinions.

Less than two months after the injunction order, and on January 6, 1881, the People's Company filed its answer, acknowledging that letters patent had been issued to Bell, as alleged in the complaint, for an alleged invention in telegraphy. However, multiple denials followed the single acknowledgment. And, unlike many legal answers, People's answers were detailed greatly.

People's denied that the alleged inventions introduced into public use by Bell were the same in construction and operation as those described in the Bell patent; denied that the patents on which the suit was brought were good and valid patents; denied that the complainant's claims had been generally accepted, but rather that Elisha Gray of Chicago, Thomas A. Edison of Menlo Park, and Daniel Drawbaugh of Eberly's Mills claimed each for himself to be the original inventor.

Further answers claimed that Bell's patent was issued unlawfully to him because no notice was given to Gray who had a caveat on file; that interferences still were pending, undetermined, in the Patent Office between Bell, Gray, Edison, and others to determine who was the original inventor of the matters described; that complainant was so dubious of its alleged rights that it permitted Gray and Edison and their licensees, to make, sell, and put into use infringing telephones and bringing no bona fide suit; that an agreement existed among Bell, Edison, and Gray providing that in the advent of a patent decision adverse to Bell the other partners would license the Bell telephones.

Additional answers said the instruments described in the first and second Bell Patents were inoperative and useless devices; and because of the state of the art at the time of the issue of patents to Bell, the claims allowed were too broad.

In further answer, People's claimed that Daniel Drawbaugh was the original inventor and discoverer of the art of communicating articulate speech between distant places by voltaic and magnetic electricity; that long before the alleged inventions of Bell, Edison, and Gray he had constructed and operated electric speaking telephones at Eberly's Mills, and had exhibited their successful operation to many persons; that his instruments contained the materials and features of the Bell instruments; that persons tested and used the Drawbaugh instruments between 1870 and 1874, and both subsequent and prior to the stated dates, and that, due to the poverty of Drawbaugh, and for that cause alone, he was unable to caveat or patent his invention. This section of answers provided the groundwork for much of the testimony in the lengthy litigation.

More answers said that after the tests (his own shop tests), Drawbaugh continued to improve the instruments and never acquiesced in the public use of Bell, Gray, Edison, Blake, or other telephones, or in the claims of the inventors; that Bell unjustly obtained patents for inventions which were first invented by Drawbaugh.

The answers continued with the defendant's claims that by purchase of right, title, and interest of Drawbaugh's inventions it had the full right at law and in equity to make, sell, and use these telephones without interference from American Bell, and without liability charged, and the defendant in good faith had made application so that Daniel Drawbaugh might receive his patents.

In conclusion, the defendant denied all unlawful conspiracy as charged, and further denied infringement of Bell patents Nos. 174,465 and 186, 787.

On February 7, 1881, the complainant's replication was filed.

Rule 67 of the Supreme Court of the United States in equity prescribed the procedure. Examiners were appointed to conduct hearings and take testimony.

Attorneys for the two sides involved were present with rights of questioning and cross-examining. The testimony was reduced to writing, affirmed to before the examiner, printed, and bound.

Drawbaugh and his attorneys spent much time visiting those who might be helpful as witnesses, interviewing them, and obtaining sworn statements. Squires Bowman of Camp Hill and Fenn of Marysville, and Constable A. B. Shank of Lower Allen Township, scoured the countryside for other prospective witnesses.

American Bell and its attorneys didn't sit back to relax, but did the same. Attorney Storrow was taken around by his local shepherd, attorney John Comfort of Slate Hill, near Eberly's Mills. Comfort remains a man of mystery in the case. Only occasional references to his work appear in the official records, but the older residents of Eberly's Mills are more voluble. Also helping were Squire Foulk of Marysville, and one Harris, manager of a New Hampshire telephone exchange, who appeared on the local scene under the name of Weeks to conduct some undercover operations.

Both sides obeyed to the letter the Biblical mandate to go out into the highways and hedges and bring them in. Storrow and Comfort journeyed west to Ohio and Indiana to question prospective witnesses and arrange for their appearance.

The complainant, as court procedure prescribes, began proof-taking first. This commenced at Boston, February 16, 1881, before Examiner Charles H. Swan and consisted principally of documentary evidence, such as charters, title, etc.

The defendant began at Harrisburg, April 19, 1881, before Examiner Frederick M. Ott in Room 5 of the Bolton Hotel.

Testimony also was taken at Philadelphia, again at Harrisburg, where with the conclusion of Thomas Draper's deposition on March 2, 1882, the examiner adjourned to meet in New York City, March 4.

In New York on Saturday morning at 10:30, the attorneys met at 234 Broadway in the office of Park Benjamin, who was a mechanical engineer and solicitor of patents. He had been an associate editor of the *Scientific American* for six years and, at the time with which we are concerned, was engaged as an "expert" by the People's Telephone Company.

The Bell Company also had its "experts," Professor Charles R. Cross and Frank L. Pope.

Benjamin had spent several months at Harrisburg, both listening to testimony and making various tests of the Drawbaugh instruments. He was ready to give his testimony concerning them, and, one by one, he compared them with the Bell instruments and enumerated the similar features. One example cited in the records: Drawbaugh instrument "B" differed in only two small details from the original Bell machine. Drawbaugh's had the armature fastened to the membrane—to "float" in the magnetic field; Bell's armature was fastened to one leg of the magnet. The second difference was the provision of a tension spring to lighten the diaphragm in "B." The expert said this instrument in a closed voltaic circuit would operate as a speaker.

By early December, 1881, all of Drawbaugh's instruments had been tested at the creek workshop. Benjamin said he ran a pair of lines from a small upstairs corner room of the shop to an outbuilding some thirty feet away. To prevent possible tampering and substitution of instruments without Benjamin's knowledge, a watchman was provided.

A large window in the outbuilding overlooked a window in the shop and each operator could see what the other was doing at all times. The expert operated at the outbuilding end of the line, and the watchman-assistant in the shop. Benjamin reports the tests and results as follows. Sketches of instruments to which he refers can be found on pages 58-64.

The tumbler transmitter "F" was put in circuit with receiver "B" and a battery; speech was heard. Magneto receiver

"A" substituted for "B," speech was heard; it was both louder and clearer. The battery then was removed from the circuit and magneto instrument "C" hooked up with "another magneto instrument." "C" tested both as a receiver and transmitter, "in both cases [I] heard speech through the line distinctly." It was found that "C" worked better as a transmitter.

Instruments "I" and "C" in circuit: the expert listened at "I" and said he heard speech clearly through the line. He then listened at "C" with the same results. Instrument "A" was tested as a receiver with transmitter "F" and a battery. "[I] heard speech passing through the line clearly and distinctly." The voice could be recognized.

Magneto instruments "D" and "E" were tested in this manner: "E" with a Tisdel instrument as a transmitter—speech was heard through the line at "E" (the Tisdel instrument was one invented by Abner G. Tisdel whose instrument rights also were owned by the People's Company). "E" was used to transmit with the Tisdel as a receiver; speech was heard clearly. Instrument "D" was tested in the same manner with the same result.

Instrument "H," the claimants' most highly refined transmitter, and one of the last to be made before the litigation, was tested with "E" as a receiver, then with "D"—and again speech came through well.

Carbon transmitters "G" and "O" were next put through the paces with a Tisdel as a receiver. The results? "I heard speech through the line with great clearness and distinctness." Then were tried "L" and "M," a pair of magneto instruments. Originally they had been combination carbon microphones and magneto receivers, but Daniel decided soon after he made them that the double duty was not practical and had removed the microphone sections. Expert Benjamin tested the redesigned pair and reported good success.

After a recess when the session reconvened, the expert continued his testimony with a report on "J." This instrument was similar to, and made after, "H." It contained double sets of carbons affected simultaneously by one diaphragm, an unsound principle. When Daniel discovered this, he put aside the instrument. It was not tested.

Even the virtues of the old teacup were considered. Of course, that hadn't existed for many years, but the exhibit, Drawbaugh sketch 3, was produced and discussed (see also Drawbaugh sketches 1 and 2). The expert held that instrument "B" connected, as shown with the cup transmitter and a battery, was an operative telephone system.

His opinion concerning the tests condensed to a single statement: "All of Mr. Drawbaugh's instruments that I have tested I know to be good, full sized, practical, working, speaking electrical telephone instruments; and with regard to his instruments and forms of instruments, to which I have referred in my previous testimony, but have not tested, I consider that these, by reason of their form and construction, may fairly be included in that description."

On the following April 28, 1882, the defendant concluded its portion of testimony-taking—but just for the moment. One hundred and twenty-seven witnesses had been called to supply the proofs-in-chief.

The next move in the extended litigation was the complainant's. It began proofs-in-rebuttal on July 13, 1882, at Harrisburg. Then the action shifted to Baltimore, to Washington, to Philadelphia, and to Boston.

Professor Bell gave his testimony at Washington from February 20 to March 13, 1883. Rebuttal closed May 10, with ninety-seven witnesses called.

The defendant began depositions in surrebuttal October 23, 1883, at Harrisburg, and continued intermittently until one hundred and ninety-nine witnesses had been heard. This took until March 24, 1884.

At Philadelphia, from April 10 to May 24, 1884, fifty witnesses supplied statements for the complainant in rejoinder. Then, the defendant took its turn again, taking testimony in surrejoinder. That began at Harrisburg, June 4, and lasted ten days while thirty-eight witnesses spoke their pieces.

Once more, it was the complainant's turn. Closing proofs, in rebuttal of the defendant's surrejoinder, were taken. The sessions held at Harrisburg, and Marysville, June 23, 1884, were brief. Seventeen depositions and some papers were received. The defendant took its two last statments the same day.

At long last, the evidence was all in—as were the attorneys. The taking of the proofs had extended over a period of three years, four months, and seven days. The sifting had been exceedingly fine.

The final printed tally included:

For the complainant: 1,047 pages of exhibits; 2,493 pages of depositions; total depositions taken, 169.

For the defendant: 160 pages of exhibits; 2,746 pages of depositions; total depositions taken, 366.

Printed pages totaled 6,446. In addition, photographs, object exhibits, drawings, maps, and moving papers were in evidence.

Revised U.S. Statute 4886 provided that "any person who has invented or discovered any new and useful art . . . may obtain a patent therefor." It was under this authority that the first Bell patent had been issued, March 7, 1876.

However, under the patent law, a claim of discovery could be predated to the time of the conception of the method and apparatus on which the patent eventually would be based, providing disclosures were made to persons capable of understanding them. Bell had made such disclosures to Doctor Clarence J. Blake and Professor Charles R. Cross, indicating he had reached the conception of exactly the method of, and apparatus for, transmitting speech by electricity. These were later shown and claimed in his patent application.

The fact that actual speech was not transmitted over his device until three days after the issue of the patent detracted nothing.

In his patent application, Bell claimed: "The method of, and apparatus for, transmitting vocal or other sounds telegraphically as herein described by causing electrical undulations, similar in form to the vibrations of the air accompanying the said vocal or other sounds (substantially as set forth)."

He had said it would talk, and it did.

Some judicial minds of the day, among them Justice Bradley of the Supreme Court, accepted a Bell "conception date" of June 10, 1875. To prove prior invention, it was necessary to antedate Bell's "conception date."

Now, what had Drawbaugh's witnesses said? The records of

the proceedings show that these witnesses had much to say, some seeming to aid his cause and others seeming to hinder it.

Samuel Snell, a stonemason of nearby New Market who had been engaged by the residents of Eberly's Mills to wall up the steps of the community spring, said he had heard of the "talking machines" and decided to see them. The inventor, at that time, had his shop and home just behind Eberly's gristmill. The witness stated he went into the mill office and listened at "a small tin can with a bladder over one end of it" while Drawbaugh spoke to him from his shop "which was about a distance of between 40 and 60 feet." He told the court he had heard the words "pretty distinctly" and thought they came through a device like "B," but he could not identify the instrument definitely. Snell said Drawbaugh told him he was working on the machine, attempting to complete it. That was in 1867, but the witness could recall neither the day nor the month.

Shank, the township constable, testified he called on inventor Drawbaugh at the creek workshop in 1869, but not to listen to the "talking machines." Instead, he was trying to collect some taxes. As he put it: ". . . Daniel Drawbaugh was back with his taxes a good bit." Testimony didn't show if the taxes were collected, but Shank said he did listen to a "talking machine." Shank said he heard what was said to him from an adjoining room—he thought the transmitter might have been twenty feet away. The instruments were on the same floor of the building and he identified the tumbler transmitter "F" and the can "B" as those used. He also testified he talked and listened (about 1875) with "A" and another instrument which he could not identify, and reported he saw "C" and "I" sometime between 1870 and 1874.

Not all testimony was received in a serious vein. Witness George Freese, a tailor and former burgess of Shiremanstown, provided the comedy relief. He had made a coat for Drawbaugh at a cost of $2.50—which of course, had not been paid. This was his story:

I went to his shop and I asked how it would suit him to pay that small bill. I then told him I was after that two dollars and a half; that this was the third time I had came [sic] after the money.

Then I told him, "Dan, it seems to me you are prety hard run"; he then had, I think, this very identical thing there [indicating machine "B"]. He tried to draw my attention to this machine, and my attention was after the two dollars and a half. He pleaded poverty; he told me I should go, I don't know the length of the end of the shop, and he talked through a concern something like this and I was at the far end of it. This instrument that he had I thought was so insignificant in my estimation I thought it was hardly anything. Says he, "George, if I have this thing accomplished," . . . I think he had something else there that I was to listen to . . . I then told him this, I still had in my mind that he was drawing my attention to this kind of foolery, I was after the two dollars and a half, didn't care about his infernal machine. I then said, "Dan if you don't make a fool of yourself as you did about the faucet." Says he, "George, my fortune lies in this, but I am most too poor." Says he, "Unless I have friends that will stick to me." . . . I didn't pay attention. It was the two dollars and a half I was after; I didn't care nor understand about his machine. After he showed me this [indicating "B"] or something like it says I, "Dan, I am after the two dollars and a half." Says he, "You must be damned hard run . . . if you are so very hard run I will get it for you." "All right," says I, "that is business on the first floor."

They went to nearby Yetter's tavern where Daniel borrowed the money from Henry Yetter and paid the debt. At the shop, the witness had managed to get his mind free of the $2.50 long enough to hear the "talking machine" perform and stated, "I heard what he said through it distinctly, as though he was alongside of me when he spoke at the other end."

Tavern-keeper, butcher Yetter was called to the witness chair to fix the date of the visit. He did this readily from entries in his books. Tailor Freese finally had collected for the coat, September 17, 1869. Yetter held no belief in the miracles of electricity and broken tumblers; so to him fate closed the door. He could never be numbered with the immortals who heard. However, he did make a good character witness. "Mr. Drawbaugh's character was a good character," he said. "He was industrious and perfectly honest, but poor. He would willingly pay every man he owed, if he would have had the means. Always perfectly sober, I don't know that I ever seen [sic] him drink a drink of whiskey; chance times he would take a glass of beer, but not unless somebody bought it for him. Did not, to my recollection, spend a cent at my bar."

John Scherich, a Mechanicsburg insurance agent, testified the afternoon of May 4, 1881. He recalled, in 1869, of listening at can "B" then on the third floor of the shop, while Daniel talked over "F" in the basement. He said he later saw "B," "F," and other instruments, probably in the fall of 1874.

Samuel Nichols, who identified himself as a "civil engineer," testified that he and his son-in-law visited the shop in 1870, saw "F" and "B," and through the latter "heard and understood what he [Drawbaugh] said, all that he spoke, very distinctly." It appears, however, the witness was rattled—in more than one way. Cross-questioning dwindled his statement to having heard just two words, and the time of his visit moved ahead to 1876. It later developed that "civil engineer" should have been "stationery engineer" at the Chesapeake Nail Works.

I. B. Millard, who lived at Eberly's Mills in the home of Ezra Gilbert from 1867 to 1869, said he had heard over instrument "B" on an outdoor line some 150 feet long, while living there. (Drawbaugh described such a line as having existed in 1877, 1878, or 1879, but not in the 1860's.) Millard said he later saw instruments "A," "D," and "E," used them, and understood over them. The date? "In cold weather before the opening of the Centennial"—that meant in the winter of 1875-76.

A miller from York County, John Zacharias, told that he visited the shop between 1869 and 1874. During that time, he had worked at the nearby New Cumberland Mills as head miller. While at Drawbaugh's workshop to purchase one of his patent trams for leveling millstones, Zacharias said conversation turned to "talking machines." Receiver "B" was then in the inventor's office and tumbler "F" upstairs. Zacharias stated that he listened at "B" while Drawbaugh spoke, and that he understood his voice distinctly, and recognized it. He said the inventor stated that it would "eventually take the place of telegraphs" and that "soon as ever he got the means, he was going to have it patented."

Hieronymus ("H" for short) Geistweit testified that in

July, 1870, he saw "F" and "B" and heard clearly through the latter. Abraham May was there in August, 1870, and said he heard and understood but could not identify any of the machines.

In the past several paragraphs, we have reviewed briefly the statements of the few witnesses who claimed to have heard via Drawbaugh's "talking machines" up to and including August, 1870. There were others who had heard of the "talking machines" and some who had seen them, and their statements, too, were an important part of the defense. By the defendant's statement, receiver-transmitter "C" was in existence, in more or less crude form, in 1870. It was an improvement over "B" as a receiver, but, for a transmitter, the inventor still depended on the good old jelly tumbler "F." Some six months after the advent of the "C" instrument, "I" was completed and was improved a short time later. It turned out to be a better receiver than transmitter, so "C" was placed in circuit as the speaker and "I" as the receiver. This gave better results than the use of "F," or so it was said—but, the majority of the demonstrations continued to be given with "F" and "B" for some unexplained reason.

Daniel Drawbaugh actually had "talking machines" on his mind before the outbreak of the Civil War or at least the three following incidents would tend to establish that he did.

While in Washington in January, 1861, he met a party of students and discussed with them a machine he was working on at the time—a machine to transmit sound. A receipt for $20, paid to the Patent Office January 30, 1861, dated the visit. Naturally, it was introduced in the evidence.

Another confirmation came from Henry B. Eberly, son of mill operator Christian Eberly. He stated: "I do recollect of him [Drawbaugh] speaking about attaching an instrument to the office, by which he could hear all that was going on at the mill office without leaving his house. He made this remark on Sunday in the mill office, in company of we [sic] boys and a party that usually congregated there. This was between 1861 and 1864; that is the nearest I can come at it; it was during the war."

Another witness, W. M. Lory, recalled that Drawbaugh had

told him, in the fall of 1864, that he was making a machine to talk a distance—twenty miles—and produced a sketch of it. Lory visioned some kind of trumpet and decided that he was talking to a fool. However, he remembered enough of the incident, when in the witness chair, to make a drawing of the sketch as he recalled it (see sketch, W. M. Lory exhibit, Appendix A).

What did Daniel say of those early days? He didn't help his own cause with his statements on the Civil War days. He said he kept no written records of any of his experiments. Some rough drawings had been made to work by but none was preserved. When questioned about his early "talking machine" experiments, he could produce no original instruments of that period but explained by drawn sketches. His original teacup transmitter was portrayed in Drawbaugh Sketch 1, the first can receiver, "B," was shown in Drawbaugh Sketch 2, and the circuit arrangement of these devices appeared in Drawbaugh Sketch 3. In early 1867, the cup was replaced, he said, by a broken-bottom jelly glass designated "F." This had been fitted with a single center rod, a bladder diaphragm, and wooden top as shown in the illustration.

Soon after, but at a time never definitely established, a rebuilt "F" came into existence with a stirrup rod arrangement, metal diaphragm, wooden top, and an unbroken tumbler to hold the components. For the litigation, a duplicate was constructed and as "F reproduced" it went through the case.

The first receiver "B" was rebuilt about 1867, it was stated. The electromagnets and can were mounted closer together on the baseboard, the armature directly connected to the diaphragm with glue, and the long connecting string thus eliminated. A strap of tin placed over the can and tacked to the baseboard on each side held it in place. Years of use were claimed for it in that condition. Many witnesses identified it and swore that they heard through it.

Of his then-living daughters, Laura, Iola, Ida, and Emma, the last named appeared to be Daniel's favorite. There were close bonds between him and his "Emmy," who at the time of point was between six and seven years old. Together, they worked to make his dream come true. She talked at the teacup transmitter while he listened at the crude can receiver. Pathetically, he described those early trials: "She was pretty small, I am positive about that; I judge her by a daughter I now have, that is about her age—over six; and, I know she was old enough to speak plainly. And, I might state another reason, she would do what I would say—would not laugh at me, as some others did."

By computation, using Emmy's birth date (September 11, 1856), it thus may be determined that Daniel's teacup was talking as early as 1863. If the story be true, her childish voice was one of the first human intonations heard through an electric telephone. Or will you here apply the maxim, *"Falsus in uno, falsus in omnibus,"* as some do?

"Emmy" died of scarlet fever, and her gravestone back of the village church bears this inscription: "Emma C. Daughter of Daniel and Elcetta J. Drawbaugh. Died Nov. 18, 1871. Aged 15 yrs. 2 mo. & 7 days . . . ."

And an epitaph now undecipherable from the ravages of weather and time.

Records of the family are fragmentary, but it is indicated that later there might have been another daughter named in "Emmy's" memory.

As the experiments with "F" and "B" progressed through the years 1867-68-69, it can be assumed the quality of transmission improved. If not, then the inventor could have more profitably employed his time with mill trams, felloes, and molasses faucets.

In commenting on the performance of those instruments for that period, he said:

> [results were] sufficiently good . . . [speech could] be heard distinctly. There were even words that were whispered; you couldn't just hear the words exactly, but could hear the whispering. I did that often. Spoken words you could hear—the whole sentences—I mean words spoken out loud and not whispering many. I would have persons in the cellar reading printed matter—some advertisement or something. . . .
>
> And, I could hear the words that were read; and at other times I would go down into the cellar and read something, and coming up, they would repeat the words to me that I had read.

30. The gravestones of Emma G. ("Emmy"), Dovane T., and Ida May, three of the Drawbaugh children. Eberly's Mills Church of God Cemetery.

Now, certainly, that was an assertion of instruments abundantly fit for commercial use and developed sufficiently to produce those results in the hands of any visitor who might desire to talk or listen. But, were those statements cold, hard facts or an enthusiastic, half-real, half-subconscious recital of what their creator hoped they would do?

Harmon K. Drawbaugh, the village singing teacher and inventor's nephew, supplies an answer. He had lived most of his life at Eberly's Mills and worked off and on in his uncle's shop. He was one of the few persons who helped with the later experiments and, of course, had heard of "talking machines" in the fall and winter of 1867 and 1868—but had not seen them until about January, 1870! This is his testimony concerning that period:

(Q) Did he [Drawbaugh] talk to you, or did you talk through those machines, "F" and "B" at that time, early in 1870?

(A) No, sir, not in words, not at that time; he used to get me to make sounds at that time—sounds that he thought had more force, or more vibrations, to move the instrument clearly. I remember very distinctly of playing the violin in it.

(Q) How soon after that did he talk to you through the instruments, if at all?

(A) A very short time afterwards; there were some words that we could understand only.

(Q) Give your best idea of the time.

(A) It was not more than ten days, or two weeks; it just happened while he was experimenting with the machine; I remember that very well.

(Q) When did you go to Waynesboro, Pa., to work?

(A) In the spring of 1872.

(Q) Prior to your going to Waynesboro to work, did you talk much, or hear him talk much through the instruments?

(A) Yes, sir, I was about the shop then a great deal.

(Q) Were the instruments in such condition prior to your going to Waynesboro, that you could hear and understand all that was [said] through them, or not?

(A) Just for a little while at a time. They appeared to require a great deal of adjusting and working at—merely trials that he got me to help him.

At this point, the testimony of a few other witnesses—with

either better hearing, or imagination—should be considered.

E. B. Hoffman, who had worked in the shop in 1868, recalled having seen Drawbaugh turning a wooden top for instrument "F" at that time. He said he heard through an instrument in the fall of 1871 and "thought" he talked through another in October, 1874. His "thoughts" about a creek flood in 1873 brought a deluge of testimony which definitely was not concerned with who had heard through what or when. Referred to as the "flood-and-bridge issue," it will be dwelt upon later.

George W. Drawbaugh, another nephew of the inventor, had lived in the village and worked in the shop. His first knowledge of "talking machines" was in 1871. He stated that he and his uncle tried talking and that he could hear something; but when they changed places his uncle could not hear. Instruments "F" and "B" were in circuit for the tests.

Jesse Eicholtz, who had worked at milling in the late 1860's and at that time operated a mill on the Conodoguinet Creek at Eyster's (Oyster's) Bridge, said he visited the Drawbaugh shop in the summer of 1871 to see and hear the "talking machines." He stated that he saw "F" and "B" and heard well, but could not identify the instrument through which he heard.

On his second visit to the shop, he was accompanied by his wife, Malinda. The purpose of this call was to obtain a sleeve for his water wheel shaft, and the date was fixed by an entry in his account book as November 6, 1873. The husband and wife both testified that they heard well—even with the shop machinery in operation. Recalled to the stand for further questioning, his testimony was somewhat embarrassing for the defendant ". . . as to the first visit, I would say that the first 'talking machine' was entirely different from the second. The first consisted of a tin can and a glass tumbler. They were connected by a string; if I mistake not, it was a string instead of a wire. That is my recollection; but it has been so long since that I wouldn't positively swear that it was a string or a wire, but I think it was a string." This did not help the case of People's Telephone.

Another disturbing "string telephone" witness was David M. Ditlow. He swore that he visited Drawbaugh's shop in

1872 and saw "a machine with tin tubes at each end and a wire wrapped with string." He listened and understood.

Perhaps the statement of Harrisburg Cotton Mill's Wilson Fox provided an explanation. He said that in 1869 or 1870 Daniel had bought yarn with which to insulate his wires. With this wrapping, the witnesses may have been led to believe that they saw strings.

Herman J. Eppler, former Northern Central Railroad superintendent of engines at Marysville roundhouse, had been called by the complainant to testify upon a collateral matter. Defendant's counsel cross-examined and drew out the information that the witness and Joseph Sadler, about 1873 or 1874, visited Milltown and the shop of a man named Drawbaugh. There they saw a cup with a string fastened to a cigar box, and heard it referred to as a "speaking trumpet." But, it developed that the shop they were in was not that of inventor Drawbaugh, but of his brother who had been toying with a "string telephone."

If it could have been established that Daniel Drawbaugh himself was working with "string telephones," the fact would make some of the testimony more understandable—but he repeatedly said he never had one. This device was a simple toy consisting of two tubes or boxes connected with a string over which speaking or music could be transmitted with a fair amount of success. It was just a toy, of course, and many children, and adults too, were experimenting with them in the 1870's, and even today youngsters do the same thing. The *Harrisburg Telegraph* of March 29, 1878, had this to say:

The articles in the *Telegraph* about a week ago, giving instructions how to make a telephone, have been productive of considerable amusement to some of our readers who have made telephones according to directions, and who are now having considerable amusement with the instrument. The most successful of these is Mr. Bishop, cigar dealer on Market Street, who has established a line to the Barber shop of Henry Gebhart on Strawberry Avenue, the distance of a square. A reporter this morning visited Mr. Bishop's store, and Mr. Gebhart was signalled for by tapping on the string of the telephone. He answered immediately, and in response to a request for music, got down his cornet and played

"Hold the Fort," the notes of which resounded through the room, and were heard as plainly as if the performer were in an adjoining room. Then followed whistling and talking, all of which were distinctly heard . . . there are a number of these homemade telephones in the city, all of which afford infinite amusement to the owners.

Ira D. Kahney, laborer, told the court that in 1872 he heard and distinctly understood through the can receiver "B"; in 1873 or 1874, he heard and understood through the improved receiver "A," and in 1875 or 1876 through "D" and "E." R. K. Shireman said he heard through an instrument fastened to the wall in about 1872—but no receiver alleged to have been made before the Bell patent date was designed to be fastened to a wall. John C. Smith stated he listened at "B" in 1872 or 1873.

In the summer of 1872, Henry F. Drawbaugh, Daniel's younger brother, was preparing to overhaul the Perry County distillery which he had operated in partnership with Aaron Egolf since 1869. Apple whisky was on the fall production schedule, and during July and August he was at the creek workshop to manufacture hoist rigging and an apple grinding cylinder. Of course, he saw the "talking machines" and through "B" heard: "Harrisburg is the capital of the state, situated on the banks of the Susquehanna River." This stock transmission phrase seemed to be in vogue at the time.

For some reason, apple whisky didn't pay off and the venture closed with year's end. Henry returned home to Newville and to his avocation of glockenspiel player in the village band. In 1876, he launched a new enterprise, a private line of produce cars which operated to New York, Philadelphia, and Baltimore. Frequent layovers at Bridgeport (now Lemoyne) Junction provided opportunities for more shop visits. On one such occasion, in 1878, Daniel spoke of Edison's telephone. The "Wizard of Menlo Park" had filed for a patent on a carbon microphone the preceding year. Concerning the conversation, Henry said: ". . . once I was in the shop; he [Daniel] then told me there was a man called Edison, who lived in New Jersey . . . that had made a 'talking machine,' and that it was his [Daniel's] invention, and he was going down to see him about it." Daniel then is quoted as concluding the conversation, " 'If you had give [sic] me the money, or helped me, or got somebody to go in with me, that man would not have got [sic] my invention!' " There is no indication that Daniel ever made the visit.

William Senseman, another witness, thought he saw instruments "B," "F," "C," and "I" in 1872 or 1873 and heard some words through one. However, he could not say which one. George Natcher said he heard through an unidentified receiver in 1872.

Mrs. George Free and her sister, Mrs. George Drawbaugh, visited the shop at a time fixed as September, 1872. Mrs. Free said she heard and recognized Drawbaugh's voice but did not undertake to identify the machine. W. H. Martin and John Keefauver were there in 1872 to talk from the cellar to the garret. No other witnesses claimed a garret instrument for that period.

The visit of Henry Bayler was no sightseeing trip, but strictly business. He had been invited to the shop to look at inventions and perhaps make an investment. The measuring faucet interested him, so in June, 1873, he went. His attention soon was diverted from faucets to "talking machines." At can "B," then on a shelf, he took his position equipped with the paper cup (Drawbaugh sketch 4) with which to listen. When the session was completed, he said, "Why, Dan . . . that is virtually a talking telegraph," and advised that a patent be taken out on it. Dan gave the usual reply, "If I had the means I would. If you will advance me the means I will give you one-half interest."

Bayler remains distinct from most of the other witnesses on at least two counts: (1) He was the only one who had been given, or remembered using, the paper cup, and (2) he said that he "thought seriously of putting in $200 or $300"—but, of course, didn't.

Daniel's brother, John B. Drawbaugh, lived next door to him, had worked in the shop off and on from 1868, and, like brother Henry and nephew George, was among the last to learn anything about "talking machines." It seems that the Drawbaughs were secretive—with the Drawbaughs. But, the

Drawbaughs were willing to help each other, as testimony shows.

When the "Welcome John" sign went up on the door of Daniel's private office in late 1872, and he was permitted to help with the talking experiments, he performed as a veteran. Like a breeze, he went through the stock transmission phrase, "Harrisburg is the capital of Pennsylvania, situated on the banks of the Susquehanna River," and kept on with, "Washington City is the capital of the United States of America, and Carlisle the county seat of Cumberland County." None could say he gave too little.

Brother Dan said that he heard, and well.

Unfortunately, in the witness chair, John couldn't remember which transmitter he had used, only that "it had a mouthpiece on top." Although permitted to talk in 1872, he stated he did not listen until, "he thinks, 1873-74." Then, he could hear only snatches of sentences—a word now and then. Daniel had to adjust continually—one instrument, then the other—but that sort of talk didn't make good testimony. Finally, defense attorney Hill saved the day—or what was left of it.

(Q) Your hearing is slightly defective is it not?
(A) Yes, sir.

Michael Smyser reported an unusual demonstration in 1872 or 1873 in which the inventor's six-year-old son participated. The receiver was upstairs with Drawbaugh and Smyser, and son Harry was in the basement with the jelly tumbler. Soon the old refrain, "Don't You Want To Be a Christian While You're Young," wafted from the mustard can. The witness was tremendously impressed—heard perfectly. Either no adjustments were required that time or little Harry was the best of all.

E. R. Holsinger, the village printer, had lived in Eberly's Mills from late 1873 to late 1876, and his home and shop were next to the community spring, about "50 steps" from the inventor's house. They became fast friends and Holsinger was able to testify intelligently concerning instruments "A," "F," "B," "D," and "E," saying he had helped experiment with them. Sometime between July, 1874, and November, 1876, he had printed a business card for inventor Drawbaugh, containing mention of about a dozen of his devices—but no "talking machines." He testified that the instruments talked fairly well.

As the local correspondent for the Carlisle *Mirror,* he supplied items of local interest, including the one on Drawbaugh's electro-magnetic clock in November, 1875. After the appearance of Bell's telephone at the Centennial in the summer of 1876, news write-ups concerning it were common.

He had the chance to lift his friend into the limelight, too, in his fall release to the *Mirror,* but chose not to give away the secret. Either that, or he believed "personal mention" type items were more important. Instead, he told his readers that George Leonard had sold the mill to Dellinger and May; that George Free was tearing down the old stone barn; that Hertzler, the winemaker, had lost two of his best cows to the grim reaper; that several very fine hogs also had departed this life. It was further reported that Mr. Horst's horse had run off and smashed the wagon; that an apple picker fell from a tree and lived two hours thereafter; and that corn huskers were earning two and a half cents a shock—but no news of the invention of a telephone. It should be here noted that the word "telephone" did not come into common usage until the appearance of news articles dealing with Bell's invention. Drawbaugh and his neighbors always spoke of "talking machines." Gradually, the new word was accepted and introduced into the Eberly's Mills vocabulary.

Returning to "hearing" witnesses for 1873, it can be found that dentist John Moffitt, Harrisburg, had known the inventor before the Civil War, visited the shop when on fishing trips to the nearby Yellow Breeches Creek, but saw no "talking machines" until 1873. That year, he heard so clearly over "A," he said, that he "was surprised and looked around expecting to find Mr. Drawbaugh behind me, but did not find him."

Moffitt, and Edgar Chellis of the 99¢ "variety palace," became partners with Drawbaugh in a faucet invention in late 1878 (this has been mentioned previously in the writing) and at that time chose to invest in the faucet rather than the telephone. Why? The answer given was that "the telephone was already in use [Bell's] and that Dan would have a hard time to establish his claim on it, and that there would be more money in the faucet than in the telephone business." So the partners had invested accordingly.

Another witness who said he saw, talked, and heard, in 1873, was Jere Kohler, a one-year resident of the village. He heard talking and understood, and had seen instruments "A," "B," "C," and "F." His trouble was that he saw too many and could not recall for the court which he had used.

On November 25, 1873, W. H. Zearing called at the machine shop to have some steelyards repaired. While this was being done, he waited in the office and stated that from a receiver on the table he heard the phrase, "The steelyards are finished." How he could have clearly heard from a distance remained an unsolved mystery.

James Brooks, who had known the inventor all his life, first saw and heard "talking machines" in the winter of 1873-74. He listened at "B" while Harry, the inventor's small son, did the talking.

Thomas Draper asserted that he was in the shop, possibly, in May, 1874. The purpose of the visit was to order an hydraulic ram for his farm at Marysville. This property adjoined the Chestnut Grove Cemetery on the west, and was under lease to farmer, George Kissinger. Mrs. Kissinger complained that the spring was too far from the house for her to carry water. Draper was a man of considerable means, and quite possibly had once owned other land there as he spoke of "Miller . . . [who] carried the chain for me when I laid out the borough of Marysville." The conversation turned from rams to "talking machines" and he was invited to become an investor. This was politely refused (a common experience for Daniel) on the grounds of being in business in faraway Delaware. However, he was willing to have a demonstration and later testified that through instrument "I" he heard "a fairy voice—kind of a feminine voice—something like a whisper . . . could understand distinctly, every word he [Drawbaugh] said; but I could not have told that it was the same person talking." The witness forgot about "talking machines" until sometime later when he heard one in Philadelphia and then thought that Drawbaugh had finally gotten someone named Bell in with him!

The complainant didn't accept the statement of an 1874 shop visit date, but rather tried to prove it as 1877. Heaven and earth almost were moved, and, when the evidence closed, each side still claimed its contention was proved.

While the reader might think that undue emphasis was placed too often on seemingly trivial matters, it must be remembered that much was at stake. A magazine article of the day gave the amount in controversy in the various then-pending telephone suits as $100 million. Obviously, American Bell couldn't afford to lose one case and perhaps see the patent go with it. The battling claimants couldn't afford to let Bell win and continue the telephone monopoly until expiration of the basic patent in 1893. To better acquaint the reader with the "Draper-Kissinger ram issue" phase of the case, it will be discussed later.

Farmer H. B. Musser said that when he had his mowing machine repaired on June 27, 1874, he heard through "B." Again the mower was fixed in June, 1876, at which time he heard through "D" and "E." W. H. Decker was positive he heard through "D" in August, 1874—but no one claimed the existence of that instrument before January, 1875.

Abraham Ditlow said he worked in the shop, and helped with the speaking experiments in the winter of 1874-75, but his testimony amounted to little. He had been brought east from Indiana to testify and had a three-month vacation in Harrisburg, with pay, while waiting to be put on the stand. Complainant's attorneys made sure that this got into the record.

The best established date for the completion of the improved receiver "A" was November, 1874. Witness Urias Nichols said he was in the shop but once in his life—January, 1875. He said he saw "B" and "F" and heard through the then, newly completed "A." By his testimony, there were, at that time, two lines from the garret to the cellar, both equipped with instruments. Son Harry, then seven years old, did the speaking from below. His testimony carried no weight since the complainant proved that the visit was not until three years later—February, 1878.

The village blacksmith, Daniel Fettrow, lived next door to

Drawbaugh for eight years and had bought his house in late 1875. During that time, he supplied the inventor with various items: hardware, smithing service, foodstuffs, fuel, and lumber to the total of some six hundred dollars which was settled periodically.

Of course, he too had heard the common gossip concerning "talking machines," but did not actually get a demonstration until March, 1875. That month, he testified, he listened at a receiver which he later identified as "D" or "E." These instruments, similar in outward appearance, had then been in existence one or two months. His son-in-law, George May, also heard through the same apparatus at that time.

A neighbor, Jacob Reneker, had his clock repaired in May, 1875, and stopped later to hear the "talking machines." As a paying patron of the shop's services, special treatment was accorded him. As he listened at what he thought was instrument "A," Daniel sang one verse of "Old Folks at Home." Most of the visitors got nothing more than talk.

C. E. Updegraff stopped fishing in the creek long enough to pay a visit and hear, over an unidentified instrument, in the same month. Orlando B. Kahney testified to hearing in August of the same year.

Testimony shows more monetary refusals were handed Daniel from time to time.

Elias Graybill, partner of the Zacharias and Company warehouse and coal yard at White Hill Station, had the opportunity to become a partner in "talking machines," too, even before he heard, but passed it up. In the early summer of 1875, the inventor, with instrument "A" in his hand, walked up to him and said that if he had a hundred and fifty dollars he could get it patented. The witness stated that he had his hand in his pocket—but hesitated. He later wished he hadn't, as the warehouse failed the following year and then as for money had "none in the bank at Mechanicsburg, and very little in my pocket." However, he did visit the shop in late 1875 and heard talking and whistling over the wires—but still, couldn't get his hand into his pocket—far enough.

Witness Jacob Evans also had trouble making up his mind —but along different lines. When called by the complainant, he swore that he knew of no "talking machines" before 1877. When the defendant put him on the stand, he recalled that he heard through an instrument in December, 1875. He could not identify it. John Simmons testified that he was in the shop in November, 1875, with Jacob Evans and talked through "A" and "E" but was confused, no doubt, as "A" was used as a receiver.

H. L. Hamme, an eleventh-hour witness called in the fourth year of the case, said he visited Drawbaugh in February or March of 1876 with two other men and heard through an instrument which he could not identify. The last witness to be considered here is David Cowens. The best that he could do was place his visit as the spring of 1876. The instrument? He didn't see any but rather had the unique experience of hearing through a hole in the wall!

His statement concluded the testimony of those who said they heard (more than forty in number) for the period 1867 to a time in 1876, after the issue of Bell's March 7 patent.

Many other witnesses also testified of hearing at later dates, of seeing various instruments, and to conversations with the inventor concerning them. Much of this evidence is discussed by Justice Wallace in his decision which will be given in full later.

i

As indicated earlier, much ado was made about some seemingly unimportant and unrelated issues. They did, however, have meaning to the case in serving to point up the reliability of some of the witnesses and their testimony and—they proved interesting. One such issue was a matter of whether or not bridges were washed out during a certain flood. This created considerable controversy.

Witness E. B. Hoffman had testified about "talking machines" for 1873 and other years, and among other things said in direct examination:

(Q) Do you recollect a freshet that occurred in that locality some years after the time you have referred to which washed away several bridges on the Yellow Breeches Creek?

(A) I do; there was a freshet in 1873, in the latter part of the summer. I can't remember the exact time. I was on the creek at that time. It washed away the foot-bridge at the mill known as Ditler's Mill at that time, and also the red bridge up at Heck's.

These statements were attacked by counsel for Bell, who expended efforts to show that the events had not occurred, reasoning that if the witness were as far off the facts on his "talking machines" testimony as they hoped to prove him on his flood and bridges story, his account would be worth considerably less than its apparent face value.

Samuel N. Emminger, clerk to the county commissioners, was brought to the stand for the purpose of setting everyone straight.

He informed all concerned that there was a red bridge at Hecks, but that it was not on the Yellow Breeches Creek, but on the Conodoguinet, some six miles away. Furthermore, the bridge had not been washed away in 1873, but was carried away by ice in 1881, he said. He also stated the first bridge over the Yellow Breeches, above Drawbaugh's shop, was Hake's Bridge. It had been built before 1860, was not red, but whitewashed, and had never been carried away or rebuilt. This proved witness Hoffman wrong on one bridge.

The other he had referred to was a footbridge at what was then known as the Etter and Shanklin Mill, a short distance below Drawbaugh's shop. Quite possibly in the past, it had been referred to as Ditler's or Ditlow's as the Ditlow farm was situated there. Emminger said this bridge had been carried away by ice and high water in the spring of 1875. As to its damage in 1873, it was testified that the walk on the York County side had washed away, not the bridge as witness Hoffman claimed. Repairs were made in less than one day, according to Henry Magonell who had helped with the work.

According to the complainant, this gave an accurate picture of just what occurred in the 1873 freshet. Nevertheless, the defendant still was able to produce witnesses who swore to the destruction not only of the Etter and Shanklin Bridge, but included for good measure the ones up-creek at the fullingmill, the clovermill, and Liberty Forge. Some forty persons, who might best be described as casual bystanders, testified to destruction, or tried to help those who did.

In answer, Bell brought in witnesses whose business it was

to know. Supervisors whose duties included the maintenance of the bridges produced their account books, carpenters and workmen who helped with bridge repairs told their stories, landowners gave statements.

Many points were brought out by each side in this issue—points each side contended substantiated firmly its case. The flood of 1873 came to be referred to as the "Watermelon Flood" as many persons down-creek from the village of Lisburn had gotten free watermelons from the unwilling donors, farmers Sipe and Wire, who had glumly watched their Kunkel's Island crop float away.

Bell concluded that its witnesses factually contradicted all phases of Hoffman's testimony. The defendant claimed that instead of Hoffman being contradicted, each of his particulars was corroborated "overwhelmingly."

Testimony of clerk Emminger was noted by Bell as being official since he reported on county records—facts. The defendant said that Emminger could give testimony about county bridges only, saying there were many smaller bridges—township bridges—which did not come under the county's jurisdiction and concerning which the county would have no record.

Conflicting though the bridge testimony had been, all were in agreement that it rained. The Monday evening, August 18, Harrisburg *Daily Telegraph* stated that within a week six and a quarter inches of rain fell in the city. The average yearly fall was from eleven to twelve inches.

One voice in the wilderness cried a warning, but none had heeded. Prophesied farmer, limekiln owner John Hursh, as he exhorted his hired hand to greater haste in gathering the oats, "We're going to have a flood—there were five moons this month." Concluded the hireling, "and the flood came, too."

It had required almost six months to hear all of the "flood and bridge" statements and, at the conclusion, rather than the lengthy testimony moving one or the other side, it made each side stand firm on its case.

ii

The seemingly unimportant side issue of the case—the Draper-Kissinger ram issue—began harmlessly enough with witness Thomas Draper's testimony concerning the date of installation of an hydraulic ram on his Marysville farm and snowballed into the major topic of discussion in and around Marysville and into as controversial an issue as that which had brought on the original suit. Some even forgot telephones and were ready to believe that the litigation concerned merely a large and very important claim for defective plumbing. Draper's testimony which precipitated the holocaust was as follows:

(Q) I understand you to say that the visit to Drawbaugh's shop you have testified about was for the purpose of ordering a ram, but that it was the following season when the ram actually was delivered and put up; is that your meaning?

(A) So far as my visit to Drawbaugh's shop is concerned, that is correct; but I am not certain that I said positively that the ram was not put in until the next season. I am not certain about that, but I am of the opinion that it was not put in until the next season. I know that there was considerable delay about getting the wooden piping bored, and I wrote to him several times about it; but Kissinger didn't get it in until sometime in the following season. I was anxious to see how the ram would work. Kissinger's excuse to me was that he couldn't get an auger to bore the pipes with. I think it was sometime late in the season. I think it was in the fall of 1875, sometime when he put the ram in; that is my impression.

He thus alleged that the visit to the shop was in 1874 and the ram installed in 1875. The device was needed to bring water from the spring to the house, a distance of 534 feet. From the spring to the ram, 161 feet of bored wooden pipes and 51 feet of metal ones were laid. From the ram to the end of its discharge pipe, near the house, 278 feet of half-inch iron pipes were placed. From the discharge point to the house was some 44 feet, the distance necessary to carry the water.

Now it was amply proven and substantially admitted on both sides that the Draper farm was located within the borough of Marysville, Perry County, Pennsylvania, lay along Valley Road, and that from April 1, 1874, to March, 1880, George Kissinger lived on the farm as a tenant; that landlord Draper and tenant Kissinger went to Eberly's Mills to order from Daniel Drawbaugh an hydraulic ram to be installed on the farm; that preparatory to the putting in of the ram, timber was cut and bored on the farm for use as piping between the spring and the ram; that iron piping also was procured to convey the water from the ram to the house; that trenches were dug for the pipes; and that, finally, the ram was delivered, installed, and put into operation. It also was admitted that there never was a ram on the Draper farm prior to the ordering of one by Draper and Kissinger of Drawbaugh, and further that no ram ever was procured for that farm except from inventor Drawbaugh. The bone of contention was the time that all these things had happened.

The defendant's theory was that the ram was ordered in the summer or fall of 1874; that the timber for the wooden piping was cut, bored, and laid in the years 1875 and 1876, commencing in the winter and spring of 1875; that old metal pipes for use between the ram and the house were procured sometime between the spring of 1875 and September, 1876; that it was installed by Drawbaugh before September of 1876; that it remained in use until the fall of 1878, giving very little satisfaction—sometimes working and sometimes not. It was claimed that it then was removed, taken to the inventor's shop, cleaned, and reconditioned, after which it was brought back, installed in the pit which then was stoned, and new iron pipes laid between the machine and the house, replacing the old, rusted, leaky ones.

The complainant sought to prove its theory on these premises: that the ram was not ordered until the fall of 1877; that the timber for the wooden piping was not cut and bored until the winter of 1877-78 and spring of 1878; that no metal piping was procured until July, 1878; and that the ram was not installed by Drawbaugh until November, 1878.

Draper in his testimony had stated that when at the shop he had heard through instrument "I." This would have been two years before the Bell patent—if the 1874 visit date could not be shaken—and another link in the chain of proof to establish the inventor's claim.

The ball began rolling with the Heisley family's testimony, for the defendant, that they bored pipe for their own spring with a certain auger in 1874; that they again used it in March,

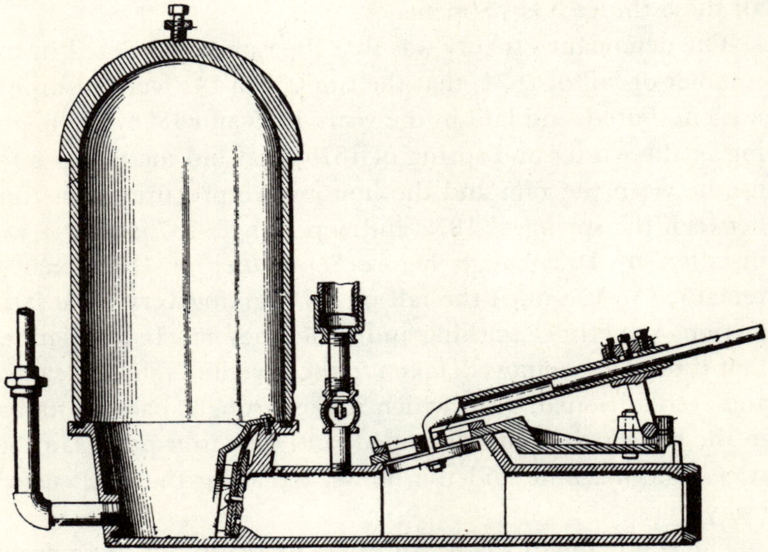

31. Not of the animal, but of the mechanical kingdom, this ram is typical of the one that created more hullabaloo than forty telephones as it went through 1,500 pages of testimony. Its original cost installed —$60.

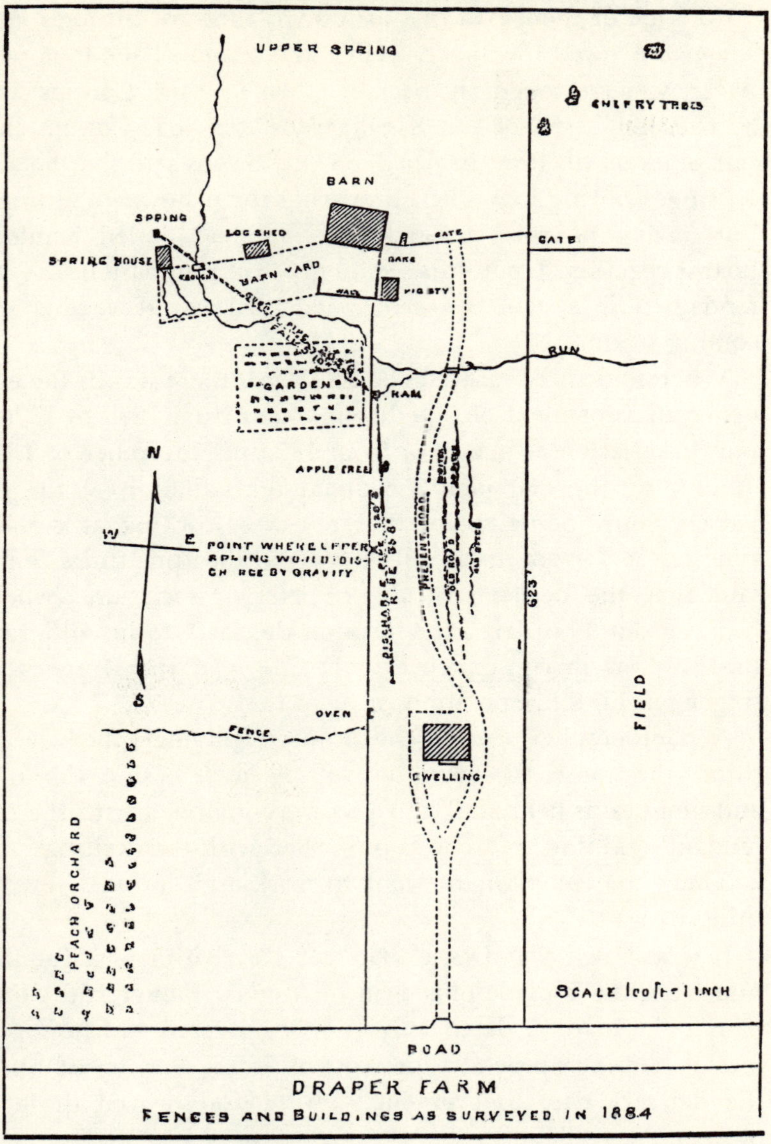

32. Here, both sides proved who did what, when—or did they?

1875. One of them said that as soon as they got through, Kissinger borrowed the auger. Keller, the owner of the tool, who never seemed to get it back, confirmed this. Cowens and Swartz, neighbors of the Kissingers, stated that Cowens had put pipes on his farm in April, 1875; Cowens said that he saw Kissinger boring logs when preparing for his own pipe line—but having no suitable sticks to bore had nailed hemlock boards together. Upon this seemingly firm foundation, the defendant built a structure—which it hoped would weather the coming storm.

When some fifty witnesses had finished, there was in the evidence an assortment of pipe boring dates from 1875 to 1876—ram installation estimates of from 1876 to November of 1878 (including the defendant's claimed reinstallation of the repaired ram). Some swore the ram was working at certain times; others swore it was out of order at the same times. Some said that the decision of the cemetery association (whose burial ground bordered the farm on the east) to install a ram in 1877 was in consequence of having seen the Draper-Kissinger machine in operation prior to that date.

A photographer sent to the property photographed everything—the house, the barn, the rolling fields beside, the hills and mountains behind. The views were included with the defendant's exhibits and one is published with this writing.

Then, the complainant went to work and "proved" everything its way.

It was shown that Daniel had purchased iron pipe for the ram from the Philadelphia firm of Morris, Tasker and Company. The invoice, dated July 1, 1878, covered two hundred feet of half-inch pipe at a net cost of $6.18. But, to establish the delivery date to everyone's satisfaction proved quite a problem.

Pennsylvania Railroad's Marysville agent, Samuel G. Geib, was called by the complainant to appear with his record books. At Baltimore, in August, 1882, he testified that the shipment was received July 2, 1878. At Harrisburg, on March 21, 1884, he was back on the witness stand with his books to testify for the defendant as to the dates of shipments of household goods to Duncannon, Pennsylvania and to Wheeler, Indiana.

It was testified that on election day, November 5, 1878, Daniel personally installed the machine. Jesse Wagner, local stonemason, swore that when he stoned the walls of the ram pit the machine was there. His time book indicated that he had done the work on November 6 and 7.

So far, so good, but the defendant's pipe boring dates of 1875-76 still were unshaken. To have prepared the logs so far in advance of their installation would have been profitless, as weathering, or "wind cracking" as it was termed, would have ruined them.

It will be remembered that the Heisleys had testified that they bored pipes for their own spring in 1874, and had again used the borrowed auger in March, 1875. Bell investigator, Brooks, while interviewing ram witnesses, stumbled upon information which proved of inestimable value. Eighteen-year-old Charles Heisley supplied the tip.

He remembered that before the second boring, his father had blacksmith William Patterson make a reamer. When the complainant opened proofs-in-rejoinder at Philadelphia, April 10, 1884, Patterson was put on the stand with his account books. He found that he had straightened the bit of the boring tool and made the reamer, from a piece of old saw blade, on March 22, 1876. His total charge for the work was forty-five cents. Defense counsel knew that much of the ram testimony would turn on that date and attorney Dixon hammered unmercifully during his lengthy cross-examination. The witness' mutilated books were termed "slovenly, inaccurate and utterly untrustworthy"—but the date held.

Eli Wise's Marysville store buzzed with the news. The word went up and down Fishing Creek Valley that the Heisleys had bored last in '76 not '75. Then, one by one, the "pipe" witnesses came back to the stand to retell their stories—with revised dates. Cowens said he relaid his pipes "a couple of years" after his original installation of 1875, and at that later date had seen Kissinger boring his logs. He was hazy as to the later date—thought it might have been as late as the spring of 1878. Then, the complainant proceeded to close the last links in its chain of evidence.

A photographer and a surveyor were hired. Theodore Ives,

at the time engaged in superintending the construction of Vanderbilt's never-to-be-completed South Pennsylvania Railroad Company bridge at Harrisburg (those old stone piers still standing in the river a few feet north of the Reading Company bridge), took his transit, two assistants and went to work on his off-day, Sunday, June 8, 1884. The Draper farm was surveyed; a map was completed in fine detail and entered in the evidence. Photographer Cotterel went to the property, June 17, to make two views to help prove that objects near the barn could be identified from the road as complainant's witness, Moore, had stated. The Cowens-Moore family farm wagon was pressed into service to haul the lensman and his paraphernalia up and down Valley Road.

Draper's wife, called to the stand, testified she was sure that on her visit to the farm in the fall of 1877 there was no ram. Daughter Virginia, a pert miss of twenty-four, said that on a visit in the fall of 1878 she noticed the installation in progress. After the testimony of wife and daughter, Draper reconsidered and wrote to attorney Storrow, of complainant's counsel, that he "feels sure the ram was not put in until 1878." When recalled by defendant, he said his impression was that it was put in either in 1877 or 1878. Kissinger himself stated that the ram had never been installed prior to the fall of 1878—and he above all was the one who should have known.

But now hear this. The time books of carpenters Taylor and Brabson showed that they were working on the Draper house in September, 1877, building a porch and an observation tower. Brabson testified that while working on the roof he quite plainly had heard the thumping of the ram while it operated. The accounts of tinsmith Strine confirmed the date of the work. His helper, Rhoads, had occasion to get water—and got it (so he said) from a pipe which came out of the ground a few feet back of the house!

While engaged in the work, Taylor and Brabson took meals at the house. At mealtimes, water was brought directly from the spring in a bucket; Kissinger was quoted as saying that the ram water was not fit to drink as it was so warm.

Strine testified he later visited the farm in November, 1877, to collect something on his bill. He found Kissinger near the barn. On the way to the house, the ram, which Strine said he noticed was not operating well, became the topic of discussion. He quotes Kissinger as saying that the ram had "been in since last fall," and that the installation had been made with second-hand pipe which leaked at the joints. At the time of his visit, the device was operating "crippled like," and delivering "a stream of water the size of a good sized rye straw" to the water barrel.

Pastor Zehner visited the farm during peach harvest time—August or September, 1878. He didn't see a ram, wasn't told they were going to get one, but was "morally positive" there was none in successful operation or he would not have heard complaints about the difficulty of getting water. He did not recall having seen pipes, logs, horses, the dry-house, outside bake-oven, or other objects proven to have been on the place at the time. About the only things he did remember seeing, and in regard to which he was more than "morally positive," were peaches, and of these he seems to have gotten a large share.

J. J. Kerr, also a man of the cloth, made several pastoral calls to the Kissinger's in 1878, but saw no hydraulic ram or pipes. But then, he had always gone in at the front door, turned to the right, and entered the parlor. It had never been contended that there was a ram installed at either of those points.

An error in the printed record of the case was reason for another summons to station-agent Geib. He was to appear at Philadelphia in May, 1884, but by then the division superintendent of the road was out of patience with the whole matter and refused to permit the record books to be taken from the Marysville station. Geib's affidavit was presented instead, but it failed to prove satisfactory.

Late Monday afternoon, June 23, Examiner Ott opened shop in the Marysville freight office to dispose finally, it was hoped, of the matter of Mr. Geib and the pipe. The books were opened for all to see, and it was established that the shipment had been received at Marysville, July 2, 1878, with freight charges of thirty-five cents due. Annexed was the release order from Daniel, dated September 6, 1878, "Mr.

Guide [sic] . . . Please let George Kissinger have the pipe at your warehouse as the pipe is for his use. [signed] Dan'l. Drawbaugh." On September 10, Kissinger sent his son-in-law, J. C. Zimmerman, with a horse and wagon, the thirty-five cents, and a note: "Mr. Gibe [sic] let Zimmerman have them [sic] pipe. [signed] George Kissinger."

The "ram issue" phase of the testimony, some 1,500 pages in length and covering every conceivable feature of the Draper water supply, from the upper spring in the hills to the water barrel back of the house, finally was finished. Both complainant and defendant were "morally positive" that they had proven their respective contentions—as "morally positive" as Pastor Zehner had been about his peaches, anyway. No one had bothered to call to the witness chair the one man who might have settled the matter very quickly—Daniel Drawbaugh.

On a sultry, summer afternoon, in the two by four office of a four by eight Perry County freight station, the testimony of the great telephone case reached an end.

The writer visited the Kissinger farm during the preparation of this work to look around a bit and to make some photographs, for he found this facet of the litigation extremely interesting. Now, he feels that he can state with authority that Mrs. Kissinger certainly needed the ram. Before its installation, she had to tote her water bucket, downhill and up, a distance of 534 feet—by a short cut possibly but 442 feet. In 1884, surveyor Ives reported that the machine appeared to be out of use, but Mrs. Kissinger had been delivered from her ordeal long before that. In the spring of 1880, the family moved to Easton, Maryland, where, it is to be hoped, water was not such a problem. George Scholl, Draper's next tenant, acquired the water worries along with his lease.

The farmhouse now is a residence, minus the observation tower, but can be identified clearly when compared with the old photograph which accompanies this writing. The barn, which played but a small part in the proceedings, has collapsed. The spring was almost dry and no traces of the ram pit could be discovered. Piped water now is supplied to the many homes of the area, and the present owner, a genial chap named Kir-

33*a*. A spring and hydraulic ram supplied this house with water in the 1870's.

33*b*. A reservoir and pipe line do it today.

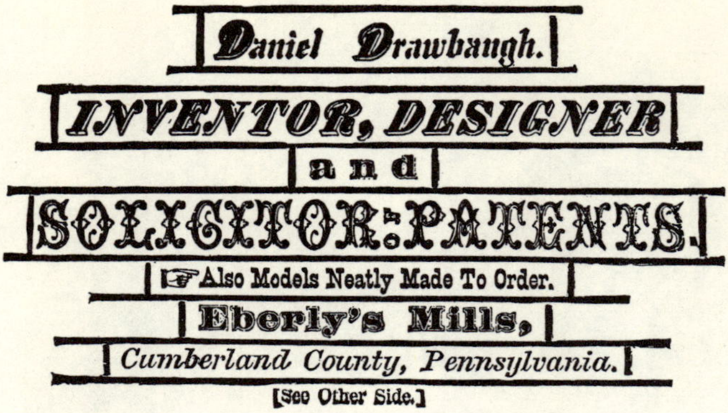

Daniel Drawbaugh.
INVENTOR, DESIGNER
and
SOLICITOR OF PATENTS.
☞ Also Models Neatly Made To Order.
Eberly's Mills,
Cumberland County, Pennsylvania.
[See Other Side.]

Dan'l Drawbaugh,
INVENTOR OF
THE FOLLOWING PATENTS.
Stave, Heading & Shingle Cutter.
Barrel Machinery.
STAVE JOINTING MACHINE, Many in use.
Tram & Red-staff for leveling face of Millstone.
Rine and Driver for running Millstone.
Nail Machinery for Feeding Nail Plates.
PUMPS, ROTARAY & OTHERS.
Hydraulic Ram.
THE DRAWBAUGH Rotary Measureing Faucet. very extensivly used.
CARPET RAG LOOPER-- A little device by which rags are looped quick and firm without Needle or Thread.
ELLECTRIC CLOCK.
MAGNETO ELECTRIC MACHINE,
For short line Telegraphing. Fire Alarm. and Propelling Electric Clocks. It can be applied to any form of Electric movement. Gives entire satisfaction USEING NO GALVANIC BATTERY.
☞ For SIMPLICITY it has NO RIVAL

34. The 1874-76 advertising card that came back to haunt Daniel during the testimony days. Many inventions—no telephone claimed. Patents for others—for him? As Kipling's Mulvaney says: "... can thim that helps others help themselves?"

wan, knows of the hardships of the previous occupants only through the author's revelations.

It will be observed that, in the present-day view of the house, the mountains seem considerably lower—but that is due to camera elevation and lens focal length which can juggle perspective a bit.

iii

The performance of Daniel's instruments at the official tests was of vital importance to the case. For that reason, an account of just what did happen has been held for discussion at this point. The report of the Eberly's Mills shop tests by defendant's expert, Park Benjamin, along with his testimony as to the results, has been related. In closing, he said he knew all of the instruments tested were good, full-sized, practical, working, speaking instruments.

Counsel for Bell said, in effect—"show us." A test day was arranged for March 28, 1882, at 11 A.M. at the office of the People's Telephone Company. At the appointed time, the attorneys convened with their experts.

The rooms between which the test line was run were located in the front and back of the building, a distance of some 84 feet. The front room was used as a receiving room; the rear room for transmitting purposes. A double line of wire was led between the rooms and into a closet in the front room. Most of the tests were made on this line. A second line was strung for reporting the results and verifying. This second line was equipped with Bell instruments and termed the "tell-tale line." A long hall separated the rooms and, with the doors closed, it was impossible to hear speech through the air. In the front room, the person receiving shut himself in a closet to cut off all street noise.

Through the three days of testing, the defendant's expert did most of the listening, but the Bell men also listened frequently.

Test words spoken into the transmitter were stenographically recorded. In his closet at the receiving end, the operator repeated what he heard to another stenographer. This was sent back over the "tell-tale line" and compared with the transmission copy. The original "A,' "E," "O," and "H" instruments were tested as were the reproduced "F," "B," "C," and "I" machines.

A word of explanation seems in order. At the start of the litigation, some of the early instruments were mere relics. The original "F" transmitter was in that category; the improved version with the stirrup rod arrangement still was in existence, but had seen its best days. A reconstructed version was placed in the evidence as "F reproduced." Magneto instrument "C" had come into the case as two small boards fastened together, one provided with a hole and a wooden mouthpiece. The inventor had been permitted to reconstruct this in what was said to be its original form. A German-silver diaphragm was put on, permanent and electro-magnets installed, and the mouthpiece altered. This went into the evidence tagged, "C reproduced." Instrument "I" when produced in evidence was an empty, walnut wood box about four inches in diameter and the same depth. To restore it to its claimed original condition, magnets and a diaphragm were added. It was designated "I reproduced."

Instrument "B" too had suffered from the ravages of time. After the original "C" was made, "B" had been more or less relegated to the shop attic, according to Daniel's statement. The evidence, however, indicated use of "B" until the latter part of 1874. (At any rate, witnesses had testified to hearing through it up to that time.) Prior to the litigation, it had been in the attic and, when brought down, was minus its bladder diaphragm. The shop mice, which Daniel never had time to exterminate, had double-crossed their benefactor. A new instrument was constructed and entered in the evidence as "B reproduced." For the sake of expediency in reporting the tests, the writer has referred to the letter designation of each instrument and ignored the terminology, "reproduced." The reader will so understand.

The opportunity to reproduce his instruments definitely was a stroke of good fortune for Daniel. He had learned much since the originals were constructed; any worthwhile improvements could be added—after all, who knew what was there originally except their creator?

Counsel for Bell hadn't overlooked this advantage and took

the view that, all "instruments reproduced" were far superior to "instruments original."

During the taking of testimony, the machines were sometimes roughly handled—seventeen of them thrown together in a basket didn't make for the best care and some were damaged. It was agreed between counsel that prior to the tests all should receive a general conditioning—under the watchful eye of the examiner, of course.

Daniel went to New York to handle that matter personally. "D" developed a break in a magnet coil and could not be repaired. It was similar to "E" which was used instead. The day before the tests, all work was completed and the claimant and his expert made the final inspections.

Then came the big day!

What a jolt it must have been that morning to the defendants present to hear their expert remark: "All he expected to get was a sound, and now and then a word."

Pope, the Bell expert, said he heard the statement and Jacques confirmed Pope.

"F" was put on the line with a Tisdel for a tuneup, taking ten minutes. "B" was substituted for the Tisdel and the tests started. Drawbaugh counted and spoke into the transmitter in a very loud voice for ten minutes.

The receiving end reported: not readily understood but could hear and understand short and familiar sentences.

Throughout the tests, speaking was usually done in loud, carefully measured tones—not in a conversational manner . . . count of one to ten, then one to twenty, which Professor Wright reported that he understood fairly well.

Transmitter "H" in circuit with a Tisdel—after adjustment, it was heard fairly well. Ernest Marx spoke, but was not understood. Drawbaugh counted and spoke some sentences with only numbers partially understood.

Tisdel taken off the line and "B" substituted—now and then some phrases were heard. The listener in his closet complained of noise; anyone on the stairs, a passing carriage or horsecar disturbed him so that he could not hear. Generally, the room was quiet.

Transmitter "F" back on the line to replace "H,"—Drawbaugh adjusted and counted one to ten six times. The word came back, "all right." Talking began, but only parts of speech could be heard. Marx took a fling at transmitting. "Hello, how are you . . . does this instrument speak distinctly?" Someone told him to talk louder. Said he, "Hello, hello, we are all deaf here, now how much worse would it be if I were to shout louder?" Newspaper items were read several times with parts of sentences understood. Then more adjustment followed: of 11 sentences transmitted, 4 words were correctly understood; 107 numbers counted, 15 heard correctly; 5 sentences read, 1 came through well enough to be understood, the rest were fairly loud but unintelligible; 19 sentences transmitted, 1 understood; 32 numbers sent, 2 understood; music box played, a few notes heard; 6 sentences spoken, 8 scattered words understood.

The last sentence, "Everybody seconds the motion to adjourn for lunch," was understood well enough to be acted on.

After lunch, tests resumed with "C" and a Bell instrument —13 sentences were read and 93 words heard. "C" was removed from circuit, "I" substituted. 138 words spoken and 61 understood; 132 words spoken, 36 heard; news items picked at random were read; Longfellow had just been buried; in Scotland, an insane asylum had experimented with the policy of more freedom for the patients, only five flew the coop; comments on biographies and the Boston stock market resume were included for variety.

"C" and "I" on the line—184 words uttered, 36 heard; 35 numbers spoken, 5 heard; 120 words spoken, 9 heard; 22 numbers sent, 9 received. Instrument "I" removed and a Bell machine substituted—66 words spoken, 14 heard; 72 words spoken, 57 heard.

Bell instrument removed and "I" returned to the line—17 words spoken, 4 heard; 14 numbers counted, 3 heard. Instruments again changed. "H" and "A" in circuit—58 words spoken, 15 understood; 62 words spoken, 13 understood.

"A" on as a transmitter with a Bell magneto as a receiver, and the order reversed—20 words understood of 38. Instruments "H" and "A" next tested—53 words transmitted, 8 heard; 59 numbers sent, 39 heard.

To vary the monotony, poetry was read and then more news items—58 words sent, 8 received. Ten minutes of adjustment and a new battery had to be placed in the circuit—52 words sent, 2 received. Benjamin, at the receiver, was asked over the "tell-tale line" why he didn't repeat what he heard. He said he couldn't—because he didn't hear. Transmission resumed with 72 words spoken of which 10 were heard. A second battery was placed in the circuit and 12 words were transmitted of which 5 were received—and this completed the day's tests.

Of course, there had been much additional conversation for the purpose of adjusting and checking. But, the results as here reported indicate the approximate overall degree of efficiency of the tested instruments. The following two days of trials followed the same general pattern. Transmitter "F" proved to be an extremely sensitive, but temperamental device; the slightest jar would throw it out of kilter. During most of the tests, it rested upon a block of stone placed on a table. That didn't solve the problem. Dead one minute, blasting eardrums the next, it required continual adjustment.

Counsel for the People's Company wanted the opportunity to hear how the early Bell machines performed, so test day was set for November 15, 1882, at Boston.

The Boston tests took less than half a day. Present at 40 State Street were the examiner, and on behalf of the defendant, attorney Hill and expert Park Benjamin. For the complainant, attorneys Smith and Storrow attended with their expert, Charles R. Cross. Shorthand writers Rydall and Yerrington also were present.

A metallic circuit was arranged between the transmitting and receiving rooms which were on opposite sides of a courtyard and perhaps 50 feet apart. A "tell-tale line" also was installed, and the scene was set for the tests of Bell's instruments.

The tests began with the older models—cone-equipped, membrane instruments, covered by the first Bell patent. Instruments 53 and 54 were placed in circuit with four cells of Bunsen battery. The results tally to something like this:

WORDS TRANSMITTED	WORDS UNDERSTOOD
37	37
100	43

165	84
82	53
16	8

The tests required about forty-five minutes and there was no adjustment of instruments from beginning to end. Some line trouble had developed, and there was a substitution for one of the instruments.

Another pair of instruments, similar to the first pair, and designated as 55 and 56 were put in the circuit.

WORDS TRANSMITTED	WORDS UNDERSTOOD
44	25
75	65

Test time—eight minutes.

Instruments 51 and 52 were hooked up for tests. They were of the so-called "gallows" type—the second Bell instruments of 1875.

WORDS TRANSMITTED	WORDS UNDERSTOOD
109	49
119	87

Test time—16 minutes.

Then another pair of gallows instruments was tested—49 and 50—the first Bell machines of 1875.

WORDS TRANSMITTED	WORDS UNDERSTOOD
12	7
158	63

With that, the Boston tests were concluded.

The results of all tests should prove at least one thing for the reader—telephones have improved through the years.

As stated before, the taking of evidence was completed June 23, 1884. The court then had to decide the issue.

While testimony was being taken, the People's Company, in 1883, began to apportion territory. Rights for the New England States were sold to the newly formed People's Telephone Company of New England, capitalized for $5,000,000. A $20,000,000 syndicate incorporated under the name, Drawbaugh Telephone and Telegraph Company, began promotion of telephone companies in the south, west, and northwest. The syndicate reserved the choice, thickly populated eastern region for its own operations but, as late as 1898, the city of

Philadelphia still had no Drawbaugh telephone. Complained *The Evening Bulletin* in its issue of July 23, "The company has been charged with seeking its (municipal) privileges in order to sell them out, and the most convincing way in which it can dispose of that charge is to begin the business within a reasonable time . . ."

The Philadelphia banking house of B. K. Jamison and Company on March 10, 1884, announced that, although subscription books for the Drawbaugh stock had closed March 4, it had secured a second block which would be offered on the terms of eight dollars a share payable on subscription, and the balance of eight dollars a share payable thirty days after a favorable decision by the United States court.

The writer made two separate inquiries for information as to stock sales and the ultimate disposition of the Drawbaugh and People's companies, but could learn nothing. However, it is known that Daniel held an entitling certificate for two thousand shares of the capital stock of the Drawbaugh company and that, under date of May 28, 1894, certificate No. 198 for two hundred shares was issued to nephew George W. Drawbaugh.

As to the disposal of the locally owned blocks of People's Company stock, little was learned. It is indicated, however, that Chellis and Daniel may have unloaded a few shares occasionally. George H. Rupley of West Fairview, was working in the shop in 1881, prior to his enrollment in the University of Michigan's mechanical engineering course. His father bought him twenty-five shares for which he paid Daniel $250. Another local investor, one Henry Cohen, wanted in on the ground floor, too. Chellis accommodated him, but nothing further concerning that transaction was revealed.

Appendix B. The Drawbaugh Instrument Photographs

When Drawbaugh was in Washington to file for his telephone patents, July 21, 1880, photographs were made of the instruments that had been brought along. From the author's files of more than 1,500 historical subjects, they have been selected to be published for what he believes to be the first time.

35*a*. "F" transmitters—original and improved versions.

35*b*. "B" the improved can receiver.

35*c*. "A"—a non-polarized receiver.

35*d*. "M"—magneto machine—the reconstructed version with the carbon microphone section removed.

35*e*. Left, exterior of the "H" transmitter. Right, interior of "H."

35*f*. Exterior and interior views of the "J" transmitter.

35*g*. "N"—magneto instrument with front plate removed.

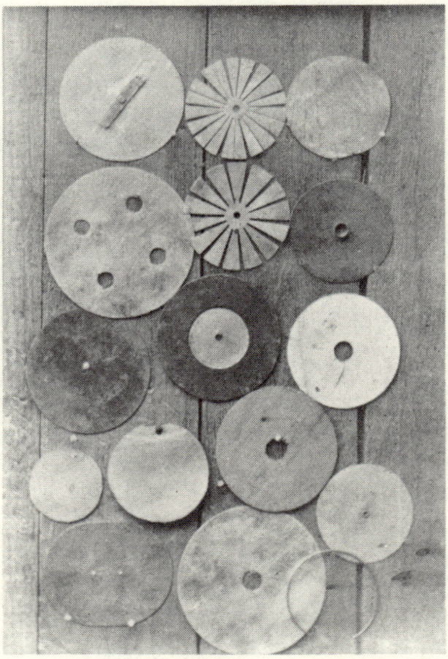

35*h*. Experimental transmitter and receiver diaphragms.

III

The Court Decisions

On Monday morning, September 22, 1884, a group of attorneys and spectators packed a small room in the New York City Post Office Building and overflowed into the hallway. Seated at the counsel tables were attorneys Roscoe Conkling, E. N. Dickerson, Chauncey Smith, J. J. Storrow, C. T. Houson, George F. Edmunds, Lysander Hill, Michael W. Jacobs, T. S. E. Dixon, and Melville Church. Perhaps somewhere in the crowd was the small, meek, blue-eyed man who was responsible for all of it—Daniel Drawbaugh. The buzz of conversation died away as a black-robed figure entered, seated himself, and tapped a gavel. With normal ceremonies, the honorable Justice William J. Wallace opened United States Circuit Court of the Southern District of New York to hear argument in the equity action of American Bell Telephone Company against People's Telephone Company.

For nine days, the pleadings continued, to close finally on October 2. Two months later, on December 2, Justice Wallace announced his decision, a full copy of which follows:

For the Complainant

36. Chauncey Smith, the veteran. Senior counsel and searcher of hearts, he planned the strategy.

37. J. J. Storrow, the good right arm. He carried it out.

38. The examiner—Frederick M. Ott. Hundreds of depositions his headache. The Philadelphia tests —a toothache.
Courtesy of the Dauphin County Historical Society

For the Respondent

39. Former Judge Lysander Hill, Drawbaugh partner and head man for the defense. His fraud charges didn't stick.
Courtesy of the Chicago Historical Society

40. Michael W. Jacobs, "Lawyer's lawyer," and Drawbaugh partner—he too had a stake.
Courtesy of the Dauphin County Historical Society

AMERICAN BELL TELEPHONE CO. and others *v.* PEOPLE'S TELEPHONE CO. and others.

(*Circuit Court, S. D. New York.* December 1, 1884.)

1. PATENTS FOR INVENTIONS—NOVELTY—PRESUMPTION FROM GRANT OF LETTERS—BURDEN OF PROOF.
   Evidence of doubtful probative force will not overthrow the presumption of novelty and originality arising from the grant of letters patent for an invention. The defense of want of novelty or originality must be made out by proof so clear and satisfactory as to remove all reasonable doubt.
2. SAME—CREDIBILITY OF WITNESS.
   Where a witness falsifies a fact in respect to which he cannot be presumed liable to mistake, courts are bound, upon principles of law, morality, and justice, to apply the maxim, *falsus in uno, falsus in omnibus.*
3. SAME—BELL TELEPHONE—DRAWBAUGH INVENTIONS.
   Upon careful examination of the testimony in this case, *held*, that Daniel Drawbaugh was not the first inventor of the electric speaking telephone, and that patent No. 174,465, for improvements in telegraphy, granted to Alexander Graham Bell, March 7, 1876, and patent No. 186,787, for improvements in electric telephony, granted to said Bell, January 30, 1877, are valid.

In Equity.

Dickerson & Dickerson, for complainants. *Edwd. N. Dickerson, Roscoe Conkling, S. J. Storrow*, and *Chauncey Smith*, of counsel.

Lysander Hill, for defendants. *Geo. F. Edmunds, Lysander Hill*, and *Church & Church*, of counsel.

WALLACE, J. This suit is brought to enjoin the defendants from using and furnishing to others for use the several inventions described in two patents granted to Alexander Graham Bell, of Salem, Massachusetts, being No. 174,465, bearing date March 7, 1876, for "Improvements in Telegraphy," and No. 186,787, bearing date January 30, 1877, for "Improvements in Electric Telephony." The issues made by the pleadings are practically resolved into the single question, to which the proofs and argument of counsel are mainly addressed, whether the patentee Bell, or Daniel Drawbaugh, of Milltown, in Cumberland county, Pennsylvania, was the first inventor of the electric speaking telephone. Concededly, Bell was an original inventor of the telephone, the principle of which, with the essential means for its application, are described in his first patent, and of the improved apparatus described in his second patent. The fifth claim of the first patent is for "the method of and apparatus for transmitting vocal or other sounds telegraphically, as herein described, by causing electrical undulations similar in form to the vibrations of the air accompanying the said vocal or other sounds, substantially as set forth." This patent has been judicially construed in two cases in the Massachusetts circuit; and in both cases it was substantially held that Bell was the discoverer of the new art of transmitting speech by electricity, and that the claim should receive the broadest interpretation to secure to the inventor, not the abstract right of sending sounds by telegraph without regard to means, but all means and processes described which are essential to the application of the principle. *American Bell Telephone Co.* v. *Spencer*, 8 FED. REP. 509; *Same* v. *Dolbear*, 15 FED. REP. 448.

In view of the conclusion reached upon the merits of the issue, it is not material whether Bell's inceptive invention did or did not antedate the time of filing his application for the first patent. That application was filed February 14, 1876. It describes apparatus which was an articulating telephone, whether Bell knew it or not. Mr. Cross, an expert, caused apparatus to be made in conformity to the description and to drawings as shown in figure 7 of the patent, which proved itself to be an operative, practical telephone. Probably the date of his inceptive invention might be carried back to July, 1875, but, irrespective of the time of the invention, the justice of his claim to be an original inventor of the telephone must remain unchallenged. It was through him also that the telephone was made known to the scientific public, and thence introduced into commercial use.

The defendants contend that long before Bell had perfected his invention, and long before its mental conception by him, Drawbaugh had not only made the same invention, but had perfected improvements in organization and detail which Bell never reached, and which were only reached years afterwards by the work of many other inventors in the same field of improvement. Their theory of the facts is stated with substantial accuracy in the answer to the bill of complaint. The answer, among other things, avers that Drawbaugh "was and is the original and first inventor and discoverer of the art of communicating articulate speech between distant places by voltaic and magneto electricity, and of the construction and operation of machines and instruments for carrying such art into practice * * *; that the said electric speaking telephones so constructed and successfully and practically used by him contained all the material and substantial parts and inventions patented" in the two patents granted to Bell, and also contained other important and valuable inventions in electric and magneto telegraphy * * *; "that some of the original machines and instruments invented, made, used, and exhibited to many others long prior to the alleged inventions of Bell are still in existence and capable of successful practical operation and use, and are identified by a large number of persons who personally tested and used and know of their practical operation and use in the years 1870, 1871, 1872, 1873, 1874 and both subsequently and prior thereto * * *; that said Drawbaugh, for more than 10 years prior to 1880, was miserably poor, in debt, with a large and helpless family dependent upon his daily labor for support, and was from such cause alone utterly unable to patent his said invention or *caveat* it, or manufacture and introduce it upon the market; and that said Drawbaugh never abandoned nor acknowledged the claims of any other person thereto, but always persisted in his claim to it, and intended to patent it as soon as he could obtain the necessary pecuniary means therefor."

Drawbaugh, in his testimony, adopts the statements of the answer as true. He also testifies that he commenced his experiments with the electric telephone as early as 1866; that prior to or as early as in 1867 he had made apparatus (in which he employed a tea-cup as the transmitter) through which speech could be transmitted feebly and incoherently; and that as early as the time of the birth of his son Charles he had so progressed that his wife, who was then confined to her bed, could, by listening with one of his instruments, hear the words spoken by him in the other instrument in a distant part of the house. His son Charles was born in 1870, and, if Drawbaugh's narrative is true, he had succeeded at that time in transmitting speech distinctly through the instruments, although whispered words would not be accurately heard. He describes instruments which he says were made by him from time to time as experiments led him from one improvement to another. He testifies that he thinks he made his first telephone apparatus prior to November, 1866, and is positive he had it before he moved his shop to the "Clover-mill" in 1867. As he describes it the body of the transmitter was a porcelain teacup, the diaphragm was of membrane, the electrodes interposed in the circuit were two copper disks, the upper one of which was connected to the diaphragm by a wire so as to vary its pressure upon a low conductor of fine earth or pulverized charcoal interposed between the disks through the action of the sound waves upon the diaphragm, and the receiver was a tin can without a top or bottom, having a membrane diaphragm stretched over one end connected by a tense cord to an armature supported on a spring and arranged close to the poles of an electro-magnet in the electric circuit. He testifies that subsequently he constructed apparatus upon the same general principle, with some change of detail, and he produces Exhibits F and B, the former a transmitter and the latter a receiver, as the remnants of the original instruments. Exhibit F is a glass tumbler; and he states that at first he used a membrane diaphragm over it, and then one of thin metal, and that for the conductor he used pulverized carbon, or carbon mixed with bronze powder, and used various tops or mouth-pieces to speak into it. The Exhibit B, he says, was the receiver, and in this he had discarded the string and the spring of his earlier receiver. He says that experiment led him to improve the transmitter, F, by substituting a metal diaphragm in place of membrane, and he produces a sketch. A reproduction of this instrument has been made by him for use in the proofs which is designated as "Exhibit F reproduced." In this the mouth-piece is modified in size and in distance from the diaphragm. He made, according to his testimony, a new receiver of more perfect construction, and produces the remnant of the original, which is designated as "Exhibit C." As he describes the instrument it was a decided advance upon the former receiver. In using this he says he tested it also as a transmitter with some success, and then improved it by placing a permanent magnet against the heel of the electro-magnet, and thus made a magneto telephone. A reproduction of such an instrument as he describes is made and referred to in the proofs as "Exhibit Reproduced C." After Exhibit C he produces Exhibits I, A, E, and D as likewise original instruments, made respectively in the chronological order of their production as exhibits. He states that I was used by him as a companion instrument to C. Exhibit A discloses a modification of form and a higher degree of mechanical adaptation. The last two, D and E, are concededly perfect, practical instruments, and according to the testimony of Mr. Benjamin, an expert witness for the defendants, would compete successfully for public patronage with any magneto telephone which had been introduced into use in 1882. It is asserted of these instruments by counsel that no higher development of the magneto telephone has been reached at the present time than is indicated by Exhibits E and D. Drawbaugh does not attempt to fix the time at which he made any of these instruments, or even the year.

He testifies, however, that he made all of them prior to the time the axle company commenced business, which was in December, 1874, except E and D which were made about that time.

The theory of the defendants is that Exhibits F and B were used by Drawbaugh in 1867, 1868, and 1869, Exhibit C in 1869 and 1870, Exhibit I in 1870 and 1871, Exhibit A in 1873 and 1874, and that Exhibits E and D were made in January or February, 1875, although cruder instruments essentially similar were made somewhat earlier. It is in proof that 33 patents were granted for improvements in telephones in 1878, 64 in 1879, more than 100 in 1880, and 94 in the first six months of 1881. According to the theory of the defendants, therefore, as early as February, 1875, Drawbaugh had not only distanced Bell in the race of invention, but also Gray and Edison, and had accomplished practically all that has since been done by a host of other inventors. The case for the defendants must stand or fall by this theory. The proofs leave no room for fair doubt that defendants' contention is substantially true, or that the defense has no foundation in fact. It is either true that Drawbaugh had long been treading his solitary path of investigation and experiment in poverty and obscurity, but had perfected his work when the inventions of other explorers were in embryo, or his story is an ingenious fabrication. And, as will hereafter appear, if the defense is a fabrication, many disinterested witnesses have contributed innocently to give it color and strength, but Drawbaugh has deliberately falsified the facts.

The complainant starts with the benefit of the presumption of law that Bell, the patentee, was the inventor of that for which the letters patent were granted him. Whoever alleges the contrary must assume the burden of proof. Evidence of doubtful probative force will not overthrow the presumption of novelty and originality arising from the grant of letters patent for an invention. It has been frequently held that the defense of want of novelty or originality must be made out by proof so clear and satisfactory as to remove all reasonable doubt. *Washburn* v. *Gould*, 3 Story, 122; *Smith* v. *Fay*, 6 Fisher, 446; *Hawes* v. *Antisdel*, 2 Bann. & Ard. 10; *Patterson* v. *Duff*, 20 FED. REP. 641; *Wood* v. *Cleveland Rolling-mill Co.* 4 Fisher, 560; *Parham* v. *American Button-hole Co.* Id. 482. In *U. S. Stamping Co.* v. *Jewett*, 18 Blatchf. 469, S. C. 7 FED. REP. 869, BLATCHFORD, J., said the defendant had not fulfilled "the necessary obligation of showing beyond any reasonable doubt" that Weber (the alleged prior inventor) was prior to Heath, (the patentee.) In *Coffin* v. *Ogden*, 18 Wall. 120, Mr. Justice SWAYNE, delivering the opinion of the court, stated the rule applicable to the defendant as follows: "The burden of proof rests upon him, and every reasonable doubt should be resolved against him." To overthrow this presumption and disprove that Bell was the first inventor, the defendants introduce the testimony of nearly 200 witnesses tending to prove the priority of invention by Drawbaugh. As the complainant concedes that Exhibits E and D are highly organized, practical telephone instruments, and fully capable of perfect articulation, the patents are invalidated if these instruments were in existence at the date of Bell's invention;

and, as will hereafter appear, either they were in existence as early as in 1875, or it is incredible that they existed at all until long after Bell's first patent had been granted and his invention had attracted general public attention at the centennial exposition and elsewhere.

In the argument for the defendants great stress is placed upon the evidence of a gradual and natural development of Drawbaugh's invention, shown by the original instruments produced, beginning with Exhibit B, and ending with the perfect magnetos E and D. It is strenuously urged that these exhibits fortify his testimony describing the instruments no longer extant, and mark the origin and culmination, beginning with the cup machine and Exhibit F, of two separate lines of invention, one leading to the battery telephone, in which the undulatory vibrations are controlled by variations in the resistance of the circuit; and the other to the magneto telephone, in which the vibrations are created in the act of producing the current itself. The general theory of the defense is substantiated by three classes of witnesses: those who heard of the existence of Drawbaugh's "talking-machines" at various times; those who talked through the machines on various occasions, or heard others talk through them; and those who attempt to identify one or more of the exhibits as the instruments they saw used. Only an outline of their testimony will be given.

More than 50 witnesses testify to having heard of the talking-machines prior to February 14, 1879.

Of these witnesses three think they heard of them in 1869; three in 1870; two in 1871; five in 1872; three in 1873; three prior to 1873; eight in 1874; two in 1875; from 1866 to 1876, one; from 1868 to 1871, one; from 1868 to 1873, one; from 1869 to 1870, one; from 1869 to 1876, one; from 1871 to 1872, two; from 1872 to 1873, one; from 1873 to 1874, one; from 1873 to 1875, three; from 1874 to 1875, one; from 1874 to 1876, one; from 1872 to 1876, one; prior to 1869, one; prior to 1872, two; prior to 1875, one.

Sixty witnesses do not attempt to identify any particular instrument, but testify that they saw a talking-machine, or talked through it or heard it talked through, at Drawbaugh's shop on occasions subsequent to 1867, and most of them fix the occasion as prior to 1876. The substance of the testimony of some of them will be given. Wilson G. Fox testified that he saw the talking-machine at Mr. Drawbaugh's shop about the year 1867 or 1868, when the old faucet company was in operation there. Prior to March, 1871, the witness was employed in the carding room of the Harrisburg Cotton-mill, and Drawbaugh came there to get material to wrap his wire to use for the talking-machine. Henry Bonholtzer testified that he was at Drawbaugh's shop in 1869, and saw talking-machines there. Margaret Brenneman testified that she saw the talking-machines at Drawbaugh's shop in 1869. Abraham May testified that he did work on Daniel Hart's house, at Milltown, in August and September, 1870, of which he produces his account-books; that he never did any work for Daniel Hart after that; that, while doing that work, he was at Drawbaugh's shop to get a boring-machine mended which he was using in the work, and Drawbaugh showed him his talking-machines, and talked through them from one floor of the shop to another. The witness understood and heard through the machine the words that Mr. Drawbaugh spoke into it. His testimony is corroborated by Jacob H. Kilmore, William H. Martin, and John A. Smith. Cyrus Orris testified that he saw Drawbaugh's talking-machines at different times from about the first of April, 1871, down to 1880, and took his son-in-law, Jacob E. Smith, to Drawbaugh's shop to see the machines about April 1, 1871. Benjamin K. Goodyear testified that in 1871 he seized the personal property of George W. Kissinger, of Milltown, upon an execution issued November 13, 1871; that on December 4, 1871, the attached property was appraised, and on that day witness went to the workshop of Daniel Drawbaugh to find J. B. Drawbaugh, to summon him as an appraiser, and had to wait for him there a short time; that, while waiting there, Daniel Drawbaugh showed him his talking-machines and talked through them to him, and witness heard him speak and understood distinctly the words that he spoke through the instrument; and that he was never in Drawbaugh's shop afterwards, so far as he can recollect. George Natcher testifies that he lived at Milltown in 1871, 1872, and never has been in the town since August 9, 1872; that while living there he was at Drawbaugh's shop, and saw and talked through the talking-machine on different floors, and listened at the same machine and understood what was said through it. Mrs. B. B. Spangler, a sister of George Natcher, testifies that she moved away from Milltown in 1872, and never has been there since; that she talked into Drawbaugh's talking-machines while she lived there; and that she was so small that Harman Drawbaugh had to lift her up to enable her to talk into the machine. Mrs. Mary Free testifies that she was with her sister, Mrs. Lydia Drawbaugh, at Drawbaugh's shop, in September, 1872, when he talked through the machines to them, and she remembers hearing through the machines, "Good afternoon, ladies!" Drawbaugh told them that the machines operated by electricity. Mrs. Lydia Drawbaugh testifies that she saw the talking-machines in September, 1872, her sister, Mrs. George Free, being present. David M. Ditlow testifies that he saw Drawbaugh's talking-machine about 1872, when Drawbaugh talked through it, and witness heard and understood through the machine what he said. David K. Ernst testifies that he was at Drawbaugh's shop with John B. Bloser about the middle of June, 1872, and talked with Drawbaugh about the talking-machines, and thinks he saw them at that time. This testimony is corroborated by John Bloser. N. W. Kahney testifies that he saw the talking-machines about 1872. William H. Martin testifies that he was at Milltown with John Keefauver, to get George Hosler to make him a pair of boots. Hosler lived at Milltown only from March, 1872, to

March, 1873. At that time witness and Mr. Keefauver went down to Drawbaugh's shop and talked through the talking-machine from the basement to the attic, and heard and understood what was said through the machines. They talked and listened at the same instrument. John F. Keefauver corroborates Mr. Martin and also states that he talked through Drawbaugh's talking-machine with Jacob M. Sadler in April, 1873, prior to the death of George B. Heck, and that about two or three years before he saw the talking-machines he had heard a good deal about them, and first heard of them at a place seven miles west of Carlisle. William W. Snyder testifies that he was at Drawbaugh's shop on Wednesday, February 5, 1873, and saw the talking-machines. He verifies the date by an entry in his diary. Jacob Barber testifies that he was a candidate for the office of county commissioner of Cumberland county in the summer of 1873, and in connection with his canvass went to Drawbaugh's shortly after the death of George B. Heck. While at Drawbaugh's shop he saw the talking-machine, and was never in the shop after July or August, 1873. Ezekiel Worley testifies that about the year 1873 he saw the talking-machines at Drawbaugh's shop. His statement is corroborated by John K. Taylor. Abraham Ditlow testifies that he knew of Drawbaugh's talking-machine in 1874, and saw it and talked through it at that time. He had forgotten the fact, but was reminded of it by Mr. Alexander Milner, of Porter county, Indiana, whom witness told about it in May or June, 1876, in Indiana. William Eppley testifies that he visited Drawbaugh's shop for the last time in May or June, 1875; that he was there several times during the two years preceding that period, and had seen talking-machines. Jonathan Fry testifies that he was at Drawbaugh's shop with Mr. Hamme and Mr. Frederick in the winter of 1875–76, and saw the talking-machines there. Jacob Evans testifies that he was at Drawbaugh's shop with his wife, his brother Andrew, and his sisters, Margaret and Sarah, about December 1, 1875, and saw and talked through the talking-machines. Henry L. Hamme testifies that he was at Drawbaugh's shop either in the last of January or the beginning of February, 1876, in company with George Frederick and Jonathan Fry, and saw and talked through the talking-machine at the time; that he heard and understood very plainly what was said through the machine even when Mr. Drawbaugh talked in a whisper. George Frederick testifies that he was at Drawbaugh's shop with Mr. Hamme and Mr. Fry in January or February, 1876, and saw the talking-machine. S. S. Rupp testifies that he was at Drawbaugh's shop with Mr. Hammacher and his scholars on February 1, 1876, and recollects that Mr. Drawbaugh at that time spoke about a machine that he had which he called a talking-machine, but the witness was interested in other things and did not pay much attention to it. George H. Bowman testifies that he saw talking-machines in Drawbaugh's shop in February, 1876, at which time somebody was talking to Mr. Drawbaugh through them.

Charles L. Drawbaugh testifies that he saw and talked through the talking-machines at Drawbaugh's shop a year or more prior to May 1, 1876, and heard and understood what was said.

The third class of witnesses are those who identify more or less positively one or more of the several exhibits as the instruments used by them, or which they saw used by others, prior to March 7, 1876. Exhibits F and B are identified by the following witnesses: Brooks saw them in 1874; Smyser, in 1872; Eberly, before December, 1870; Wagner, in the fall of 1874; Freese, in 1869 or 1870; Yetter, about Christmas, 1875; Fry, spring of 1875; Carl, in 1870; Scherick, in 1869; Balsley, between 1870 and 1874; Good, before 1872; Kahney, in 1871 or 1872; Schettel, about 1872; Nichols, in 1875; Renneker, in May, 1875; Weber, late in 1874; Stephen, before 1875; Shireman, about 1872; Hawn, about 1872; H. B. Eberly, in May, 1873; J. C. Smith, between April, 1872, and April, 1873; Sternberger, in October, 1871; Fetterow, in April, 1876; Halsinger, prior to 1876; Shoop, in 1869; H. F. Drawbaugh, in 1872; Zimmerman, in 1871; Bates, in 1874; Guistweit, in July, 1870; Hale, in fall of 1873; Stone, in June, 1871; Free, in June, 1872; J. A. Oyster, in June, 1875; Harman K. Drawbaugh, in January, 1871; J. B. Drawbaugh, in 1869; G. W. Drawbaugh, in 1870; Lenseman, in July, 1871; Fisher, in 1868 or 1869; Hubler, in fall of 1873; Updegraff, in 1874; W. H. Decker, in 1873; and a number of other witnesses saw one of these two exhibits.

The identification of Exhibits C, I, and A is made by a smaller number of witnesses. Some of them think they saw C in 1870, and others at various dates after that and as late as March, 1876.

One of the witnesses thinks he saw I in 1871, the others locate the occasions in 1873, 1874, and 1875. Some of the witnesses think they saw A as early as 1872, one of them in 1870; but most of them saw it, they think, in 1875.

Exhibits E and D resemble each other very closely in appearance, and most of the witnesses produced to identify them saw both at the same time. They locate the time as follows: Fry, laborer, in May or June, 1875; Fry, farmer, in April, 1875; Bayler, in June, 1873, (Exhibit D;) Springer, after April, 1876; Schettel, about 1875; Shoop, after February, 1877; Musser, in June, 1876, (Exhibit D;) Millard, in 1875; Holsinger, in summer of 1875; Shoop, in 1874 or 1875; Bates, between 1874 and 1877; Dellinger, in March, 1876, (Exhibit E;) Gustweit, between 1870 and 1876; Bowen, in September, 1878; Hale, in fall of 1875, (Exhibit D;) Michael Dellinger, in November, 1877, (Exhibit D;) Harman K. Drawbaugh, in January, 1875, and helped put up wire for them; J. B. Drawbaugh, prior to January 26, 1875; George W. Drawbaugh identifies all the exhibits as seen by him sometime between 1871 and 1878; Updegraff and Musser, in 1876; Smith, in 1872 or 1876, (Exhibit E;) May, in 1876, (Exhibit D;) J. H. Smith, in May, 1876, (Exhibit D;) Decker, in 1874,

(Exhibit D;) Vannasdale, in February, 1875; Evans, in fall of 1875; Mrs. Erb, in fall of 1875; S. E. Evans, in fall of 1875, (Exhibit D;) M. E. Evans, in fall of 1875, (Exhibit D.)

Some of the witnesses who identify exhibits identify the whole series. Other witnesses besides those named identify one or more of the exhibits as seen by them at times subsequent to the date of Bell's application for his patent. Some of the witnesses who identify one or more of the instruments exhibited to them by Drawbaugh as the Exhibits F, B, or C, saw or used them in 1875 or 1876. Among these are the following to whose testimony a reference will be made: Mr. Springer testifies that he repeatedly talked and listened with Drawbaugh through the instruments, after the first of April, 1876, using Exhibits F and B as the instruments. Mr. Musser testifies that he talked through F and B in June, 1874, but the proofs show that this occasion was as late as in the summer of 1876. Mr. Moore, who is produced to show that Drawbaugh applied to him to acquire an interest in the invention, testifies that the talking-machine which Drawbaugh produced was Exhibit B. This was in May, 1875. Mr. Bayler testifies that he talked through F and B in 1873, but the proofs show that the occasion was between 1875 and 1877. Mr. Nichols locates the middle of January, 1875, as the time when he saw Exhibit B in use.

That the talking-machines referred to by the witnesses were electric instruments is clearly established. Drawbaugh testifies explicitly that they were always used with a closed circuit, and without breaking the current, some of them being battery telephones, and some magneto telephones. He always represented them as actuated by electricity to those to whom he explained or described them, and claimed his invention would supersede the telegraph. His assertions show them to have been electrical instruments. He stated to the witness Shank, "It was the greatest invention ever known; if he had the means to go on with it they could talk, or rather be a time to come as to talk, to the old country same as we can talk here." To Zacharias, that "he could run it out for miles, and parties could talk in at one end and be heard at the other end the same as persons in a room together." To Smith, that "parties between Harrisburg and Philadelphia could communicate as if they were speaking together; there would hardly be any limits;" it was an "instrument to convey the voice—to supply the place of the telegraph." To Smyser, that it would work "from here to California." To Fry, that one "can talk as far as the wire goes." To Carl, that "he could hear a man talk from that place to New Cumberland or Harrisburg, and understand distinctly what he said." To Sherwick, that it was "better and handier than the telegraph; that you could just talk through it in place of writing." To Balsley, that "by attaching two wires you can hear it away off; the telegraph is nowhere with it." To Kahney, that "he could talk the same for miles as he could for a short distance." To

Shettel, that "if he had a wire from the shop in connection with the telegraph wires at White Hill he could talk to Mechanicsburg by having a machine there or an instrument in the office; that it would be better than telegraphing, and that it would be worth a great deal of money." To Reneker, that "he thought he could make it that he could talk through to Harrisburg; he thought they would take the place of telegraphing." To Weber, that "it beats all the others of my inventions; he could carry sound, or rather talk, as far as Shiremanstown." To Hawn, that "he would be able to operate, that a man preaching in New York, that a congregation in Philadelphia would hear the same sermon." To Kahney, that "he could just as easy speak ten miles as one, or any distance he would choose to." To Rupp, who was there with Hamacher, that "it was worked by electricity; would take the place of the telegraph, and that he could make it so that he could talk to San Francisco." To Musser, that "he was going to make a machine to talk from Harrisburg to Philadelphia, and it would be cheaper and quicker way than telegraphing." To Smith, that "he believed they could talk for a hundred miles." To Fetterow, that "I could speak ten, fifteen, or twenty miles, or even to California if there was a wire extended." To Wisler, that "he could attach a wire to it and talk for ten miles—as far as he could have a circuit around." To H. F. Drawbaugh, that "he could talk across the continent." To Free, that "the talking-machine could be used to talk at a long distance—from Philadelphia to California." To Landis, that "it could be used a thousand miles; it would take the place of the telegraph." To Lenig, that "he could talk hundreds of miles through that." To Updegraff, that "instead of using the old mode of telegraphing he could talk directly through the wire; he thought he could talk as far as you could use the ordinary telegraph wire." To Draper, that "he thought it was or would be one of the greatest inventions of the age, and would take the place of telegraphing." To A. Evans, that "he could take this machine and talk clear out to Europe across the ocean." To Eicholz, that "if he could only get some one to help him once he would run it to Harrisburg and convince them, and then he would run it from Harrisburg to Philadelphia." He stated to the witness Shank, that "it works by electricity." To Smith, that "it was by electricity." To Nichols, that "the sound was conducted by electricity." To C. Eberly, that the instruments were "to convey sound by electricity." To Coudry, that "they were operated by electricity." To Shoop, that "it operated by a battery." To Shireman, that "they operated by magnetism." To Hawn, that "they would be operated on by a battery." To N. W. Kahney, that "the machine was operated by electricity—by a battery." To Zimmerman, that "it was electricity that would pass it over the wires; that it would carry the sound right along." To Hale, that "it was driven by a magnet." To H. K. Drawbaugh, that "the sound could be carried to a distance on a wire by the use of electricity." To Lenig, that "electricity was used in connection with it." To Prof. Heiges, that "in connection with a talking-machine both magnetism and electricity were applied." To Goodyear, that "his talking-machine was also done by electricity over wires." To Woods, that "it was to be an electric machine used in place of telegraphing." To Young, that "it was an electric talking-machine which he had invented."

Thus Drawbaugh is corroborated by a cloud of witnesses whose testimony tends to substantiate his narrative. Without stopping at this point to consider the credibility and probative force of their testimony, it suffices to state that, although some of the witnesses seem to have been reckless and unscrupulous in their statements, the great body of them are undoubtedly honest witnesses. It is impossible, however, to believe that Drawbaugh can be mistaken in the substance of his testimony, and the conclusion cannot be ignored that either his testimony is true, in its essential parts, or his narrative has been manufactured to fit the exigencies of the case. In order to ascertain what effect is to be given to the corroborative proofs, it is important to determine whether Drawbaugh is an honest witness or whether he has intentionally falsified collateral facts, and is therefore to be deemed discredited. If the defense is to be believed, he had been experimenting with his talking-machine from 1866, and had successfully transmitted speech as early as 1870, if not before that time. He testifies that he had used Exhibits B and F in transmitting speech for two or three years before he made Exhibit C. According to the theory of the defendants, Exhibit C was made in 1869 or 1870. At that time he had reached a secondary stage in the development of his invention, and certainly as early as in 1872, when Exhibit C had received its latest modifications, the invention had passed out of the period of rudimental forms embodying principle merely, into a form embodying nice details of construction, and had reached a perfection not reached by Bell in his earlier patent. Drawbaugh was well aware of the merit and of the great pecuniary value of the invention. He had obtained patents for several inventions of minor value; yet, from 1870 until July, 1880, he did not apply for a patent for the telephone. It was of the first importance to explain the reason of his inaction, because it seems incredible that the inventor of the telephone should not only omit to patent it as soon as he could, but should also remain silent for years after others were winning the fame and profits of the invention. Only one explanation was possible, and that has been attempted. As stated in the answer and in his testimony, it is that he was unable to do so by reason of his poverty. The answer alleges "that for more than ten years prior to 1880 he was miserably poor, and utterly unable to patent his invention or *caveat* it." He was asked the question: "Do you mean to have it understood from your last answer that there was any other reason for some period prior to 1870, except your poverty, whether greater or less, which prevented you from patenting your invention or filing a *caveat* for it?" His answer was: "If I understood that right, there was no other reason that I can think of now." He proceeds to state that Exhibits F and B exhibited the principle perfectly enough to patent.

In the elaborate efforts of the defendants to substantiate the theory of Drawbaugh's inability from poverty to patent his invention, much testimony has been produced to show, and which does show, that he was always more or less in debt, often a borrower of small sums of money, was dilatory in paying his debts, and used to plead his inability when dunned, and was often sued, and judgments and executions were obtained against him; but it is clear from a few plain facts that the theory of extreme poverty is unfounded, and that Drawbaugh is dishonest in putting it forward. In 1867 and 1869, besides what he received for his wages, he received $5,000 from the pump company for his faucet invention, besides $1,000 in the stock of the concern. On the first day of April, 1869, he received $1,000 from one Gardner, for the sale of a half interest in a faucet invention. He invested $2,000 of the $5,000 in real estate, lost $400 of it in an apple speculation, and used the $1,000 received from Gardner to buy a house and lot for his father. Between 1867 and 1873 he paid $1,200 to the Drawbaugh Manufacturing Company for assessments on his stock, besides $870 in labor; and in July, 1873, received from that company $425 cash, in settlement of its affairs. From 1867 to April, 1872, he was the owner of real estate, for which he had paid $2,300 in the fall of 1867, and upon which he expended in improvements, in the spring of 1868, from $300 to $400, and which was incumbered only by a prior lien for $300. In the spring of 1872 he incumbered it for $1,000, not as a principal, but as a surety. He was in receipt of $110 annually as rent for a part of this property, occupying the rest himself until he sold it in 1876, and bought another house in the town of Mechanicsville. He was always in receipt of fair wages for his labor. From April 1, 1875, to April 1, 1876, he received nearly $450 for wages from the axle company, irrespective of his earnings from other sources, and declined steady work at times, because he could make more by job-work. Thus it appears that, although at times it was not convenient for him to pay his debts, or he was careless or indifferent, he had not only the means of raising money during all this period, but that on many occasions he had means for investment and for speculation. The pretense that he could not raise the fees to *caveat* or patent his invention is transparently absurd. He was accustomed to prepare specifications of patents, and was a maker of models, and advertised himself as an inventor, designer, and solicitor of patents. During the time he was experimenting on his talking-machine, and before he applied for a patent, he found time and materials for experimenting with and making the Giffard injector for steam-engines, the autograph telegraph, the magneto-dial telegraph, the magneto key, the automatic fire-alarm, and the electric clock. During this period he was a friend of Mr. Weaver, a patent solicitor, who frequently gave him advice and professional assistance in return for mechanical services rendered by Drawbaugh, and who drew specifications for him for a measuring faucet and for the magnetic clock. If he was not competent himself to make an application for the patent, it cannot be doubted that, with the assistance of Weaver, he could have made a proper application at a trifling outlay, if any, beyond the fees of the office.

Drawbaugh devoted a great deal of time between 1867 and 1878 to the invention and construction of his electric clock, and the time and money expended by him in experimenting and constructing this clock in its various forms, especially those made in 1877 and 1878, was much more than would have enabled him to patent his talking-machine. These clocks were built by him with his own tools and out of his own money, and, to build them economically, he made a gear cutting-machine which must have cost him more than it would to patent his telephone. In April, 1878, he received $500 from the Electric Clock Company for the privilege of using his clock invention.

In order to fortify the theory of Drawbaugh's inability from poverty to patent his invention, the defendants have attempted, by testimony from him and from others, to show that he was extremely solicitous to patent it, and tried to induce others to furnish the means. Mr. Springer testifies that "his (Drawbaugh's) whole mind appeared to be on his talking-machine; he told me that many a night he didn't sleep just studying how to improve it." After May, 1872, according to the testimony of Jacob Hawn, the talking-machine superseded the clock in Drawbaugh's interest. According to Mr. Holsinger, from 1873 to 1876 "he appeared to be crazy on it; I often tried to get information from him on other subjects, and about half a minute's talk would turn him right on the talking-machine." Henry F. Drawbaugh, his brother, testifies: "Every time I was down there, from the summer of 1872 to 1879 or 1880, he was working at it and talking, and wanted me to go in with him and furnish means." Mr. Bates says he was in Drawbaugh's shop eight or ten times between the summer of 1874 and the fall of 1877, and "his general conversation was about the talking-machine; said he would like to get it patented, but had not the means, and could make a fortune out of it." Drawbaugh testifies as follows:

"*Question.* A good many witnesses have testified that you were at various times talking of patenting your electric speaking-telephone invention: what is your recollection about that—did you intend to patent it or not? *Answer.* Yes, sir; I intended to patent it. I had spoken to a number of persons to assist me. I would state to them that I would give them an interest in the invention for them to furnish the money to have it patented. *Q.* Why did you not patent it with your own money? *A.* I didn't have any money. *Q.* At how early a time did you have the intention of patenting it? *A.* I could hardly say how early. I spoke to persons even at an early time. I spoke to

Christian Eberly; it may have been prior to 1870 I spoke to Frank Lee; I spoke to them about taking an interest. They were among the earliest. I can't remember all the persons, as I had spoken to a great many."

Lee is not a witness, having died in 1872. Christian Eberly locates the time as between 1867 and 1870. He had been a partner with Drawbaugh in a number of inventions, and was a capitalist. He was asked:

"When Mr. Drawbaugh showed you his talking-machine, state whether he proposed to you to go into partnership with him and furnish the money for that also, as you had before that time, on the other inventions?"

He answered:

"Not altogether; he intimated that he would take me in. I don't recollect as I said anything, or what I said."

The witness was often in Drawbaugh's shop subsequently, in 1871, 1872, and 1873, but mentions no other proposition. The only other persons Drawbaugh specifies as having been applied to by him are Capt. Moore, Henry Bayler, and Simon Oyster. Oyster was not called as a witness. Capt. Moore was examined as a witness for the defendants, and his testimony is significant. He was the principal of the Soldiers' Orphans' School, an institution in the vicinity of Eberly's Mills, and was the secretary and treasurer of the axle company, a concern that in part occupied Drawbaugh's shop in 1875 and 1876. He testifies that about May, 1875, Drawbaugh showed him a talking-machine; said he was unable to patent it himself, and desired witness to "go in with him and get a patent." He states that he told Drawbaugh he didn't want to go into any new inventions, but that it would be a fortune to any person bringing it out if it could be put to practical use. He identifies Exhibit B as the only machine shown him at that time by Drawbaugh. Although he and Drawbaugh maintained intimate business relations for a year after that time, the subject seems never to have been referred to again. Mr. Moore was an intelligent capitalist. It is strange that Drawbaugh should have shown him Exhibit B, one-half of the crude instrument of 1867–1869, if the perfect instruments, E and D, were in existence; and more strange that the subject was never mentioned again between them, or that no attempt was made to speak through any machine, if they had any faith in the value of the invention. Mr. Bayler, the other witness, carried on lumbering and a saw-mill from 1873 to 1877 in the vicinity of Milltown, and employed Drawbaugh frequently to repair machinery. He testifies that in June, 1873, Drawbaugh showed him the talking-machine, and he said to Drawbaugh, "Why, Dan., that is virtually a talking telegraph," and advised him to take out a patent for it, to which Drawbaugh replied: "If I had the means, I would; if you'll advance me the means to procure a patent I'll give you a half interest." The witness continues: "Generally, on him meeting me, he would urge it,—urge me to take an interest by furnishing him the means to take out a patent." He also identifies Exhibits F and B as the instruments shown him by Drawbaugh. But his books show that during all the time from April, 1873, to May, 1876, he owed Drawbaugh more than the fees necessary for procuring a patent.

The defendants produce other witnesses to prove that from 1870 to 1879 Drawbaugh was showing his telephone, adverting to his poverty, and trying to induce somebody to assist him. Mr. Herr may be cited as an illustration. He testified that in 1870 or 1871 Drawbaugh wanted money to get a *caveat* to secure his invention, and told the witness if he would help him or procure any person to assist him he would give him a half interest. Without adverting further to the testimony on this subject, it is sufficient to say, in view of the fact that there never was a time from 1867 to 1880 when Drawbaugh did not have the money to *caveat* and patent his invention, or the means of borrowing it, the only legitimate effect of such testimony is to discredit the whole defense by exciting the suspicion that it is bolstered up by exaggerated and unreliable testimony. It will hereafter be shown that among the men with whom Drawbaugh maintained business and friendly relations during this period there were many of intelligence and means. Some of them may have distrusted his judgment and regarded him as a visionary; some of them may have been indifferent or timid; but it is incredible that when only a trifling sum was required for a half interest in the invention none of them could be sufficiently impressed with its merit or financial value to investigate it seriously as a speculation or an investment. He induced persons to invest in faucet inventions and in his magnetic clock; and it cannot be true that he could find no one to entertain the talking-machine, which, according to the common rumor of the neighborhood, was to supersede the telegraph, and, in the words of one of the witnesses, "make Drawbaugh the richest man in the Cumberland valley." It was very natural that a hard-headed old farmer like William Darr, on being told by Drawbaugh that he had a machine by which he could talk across the Atlantic ocean, should advise him to "try it first in talking across the Yellow Breeches creek;" but it is beyond comprehension or belief that none of the capitalists or speculators about him could be induced to seriously consider it, if it was an operative device. Where a witness falsifies a fact in respect to which he cannot be presumed liable to mistake, courts are bound, "upon principles of law, morality, and justice, to apply the maxim, *falsus in uno, falsus in omnibus*." *The Trinidad*, 7 Wheat. 283. Drawbaugh could not be mistaken in asserting that it was his poverty which prevented him from caveating or patenting his invention. He was not led to the assertion inadvertently. Those with whom he is associated in the defense understood fully, and so did he, that the fact that a professional inventor and patentee did not go to the patent-office to secure an invention like the telephone for 10 years after it had been completed and demonstrated was almost conclusive against the theory that he had made the invention, and that, unless this presumption could be parried, no court would credit his story. The theory of constraining poverty was therefore formulated in the answer, elaborately fortified by witnesses, and testified to by Drawbaugh. It is overthrown by a few plain, indisputable facts, and Drawbaugh's veracity falls with it.

The defense must rest upon the testimony of the witnesses who corroborate Drawbaugh. The case made by these witnesses is sufficiently formidable to overcome the legal presumption of the validity of the complainant's patents. It is met by the complainant with rebutting evidence, direct and circumstantial, showing the intrinsic improbability of the theory that Drawbaugh was the inventor of the telephone, and showing his conduct or declarations inconsistent with any hypothesis that he was more than an unsuccessful experimenter with the invention. Many witnesses have also been produced by the the complainant to attack the credibility and reliability of the testimony of the defendants' witnesses. Of necessity the testimony of most of the defendants' witnesses can only be attacked by showing that the witnesses are mistaken as to the time when they saw Drawbaugh's talking-machine, or as to what they really saw on the occasions they refer to. The way in which the testimony of Uriah P. Nichols is met will illustrate the general tenor of such testimony. Mr. Nichols was one of the most intelligent and trustworthy of the defendants' witnesses, a farmer and machinist, who testified that on the eighteenth day of January, 1875, he visited Drawbaugh's shop on business, saw two instruments which he identified as Exhibits B and A, and he described their mode of operation as stated to him by Drawbaugh at the time. He says he listened at one instrument while a boy spoke into another 200 feet away, connected by wires, and heard the boy say: "Is it you, father, speaking?" The complainant produces nine witnesses to show that the occasion could not have been prior to February, 1878. The witness fixes the date by a purchase of lime made by him on the visit, and states that he went to Drawbaugh's to see an electric clock of which he had recently read a description in a newspaper, and soon after the visit told Mr. Maish and others about the telephone he had seen at Drawbaugh's. The complainant proves that the newspaper article was not published until February, 1878; that when the witness told Mr. Maish of the telephone at Drawbaugh's, the latter, who was then a member of congress, remembered the occasion, knew all about Bell's telephone at the time, and had used it in Washington. Mr. Maish states that, as Drawbaugh was one of his constituents, he would have been deeply impressed by the conversation if he had understood Drawbaugh claimed to be the inventor. Without attempting to particularize the rest of the testimony for the complainant upon this issue, it suffices to say that several other witnesses were introduced to show that the lime was not purchased by Nichols before 1876. Much testimony is given by complainant upon collateral issues of a similar character. One of these issues relates to the time when Thomas Draper ordered a hydraulic ram of Drawbaugh. Mr. Draper was an important witness for the defendants. He testified that he went with Mr. Kissinger, a tenant of his, to Drawbaugh's shop in May or the early summer of 1874, for the purpose of ordering of Drawbaugh a hydraulic ram to be used upon the farm Kissinger had leased of him in April, and that he was never at Drawbaugh's on any other occasion. He identified Exhibit C positively and Exhibit I less positively as the instruments used and through which he listened while Drawbaugh talked. The complainant proved that the hydraulic ram was not put to use until the fall of 1878, and undertook to locate the date of Draper's visit approximately by that fact. Seventy-five witnesses were introduced by the respective parties upon this collateral issue. These illustrations show how hopeless a task it would be to review the testimony satisfactorily or analyze it minutely. Five hundred witnesses have been examined by the parties upon the main question and the collateral issues, and their testimony is in a printed record of over 6,000 pages. If it were practicable to do so it would not be profitable, because a microscopic view of the controversy would be inadequate and misleading. In cases where such a chaos of oral testimony exists it is usually found that the judgment is convinced by a few leading facts and *indicia* outlined so clearly that they cannot be obscured by prevarication or the aberrations of memory. Such facts and *indicia* are found here, and they are so persuasive and cogent that the testimony of a myriad of witnesses cannot prevail against them.

The first group of facts of this nature are those which bear upon the capacity and character of Drawbaugh as an inventor, and tend to show that it is not only highly improbable but almost impossible that he could have been the author of the telephone. In the summer of 1878 he composed a biography of himself for publication in the history of Cumberland county, which presents a graphic picture of the inventor and of the man. He commences by describing himself as "born in the quiet, secluded village of Milltown, three miles from Harrisburg," and as "one of the greatest inventive geniuses of this age, who has spent the greater part of an active life conceiving and producing, as the result of the conceptions of an unusually fertile brain, a score of useful, ingenious machines and devices." "It appears," he says, "by examining a list of his inventions, that the manufacturing interests of the place in his boyhood days gave direction to his thoughts and incentive to his actions." He proceeds to enumerate a list of his inventions as follows:

"His first invention was an automatic sawing-machine; then a number of machines used in wagon-making; then a machine for boring spoke tenets; then a machine for sawing tenets; a barrel-stave jointing-machine, patented in 1851. This machine was pretty generally introduced, and its merits appreciated. An automatic grinding-machine was next invented to meet a demand created by the introduction of the jointer; then followed several machines for making stave headings and shingles, all of which were patented in 1855; after which, machines for rounding, heading, crozing, dressing, and finishing outside of barrels were invented. These were again followed by device for running mill-stones; one for dressing mill-stones; a device for elevating grain in mills. He then invented and had patented four improvements in nail-plate feeding; next a tack-machine and a new design in tacks. Photography next engaged his attention. He fitted himself for action in this field by manufacturing his own camera ground, and fitted acromatic lenses for camera, prepared the necessary chemicals, and improved the process for enlarging pictures. Next electricity and electric machinery attracted his attention, and an electric-machine was produced, throwing out of consideration the galvanic battery and electric pile; then a machine for alphabetical telegraphing; then the justly-celebrated electric clock and the machinery necessary for its construction; and several kinds of telephones: one of which is operated by battery, and another by induction."

He concludes as follows:

"It will be seen from the foregoing that Mr. Drawbaugh has penetrated vast fields in search of information, and with what success we leave it to the readers to determine. We are proud to own Mr. D. as a citizen of our township, and deem him worthy of a position at the head of the list of our prominent men, and are happy to accord him that position."

This portrait, drawn by himself, depicts, without the aid of extrinsic evidence, the ignorance and vanity of the man, and the incongruous and fantastic assortment of his inventive projects. It suggests also the character of a charlatan. That he was a skillful and ingenious mechanic is undoubtedly true. Invention was his hobby and his vocation. But that he was an inventor in a large sense is disproved by the nature and results of his work. Every patent that he obtained was for some improvement on existing devices, which involved mechanical skill rather than any high degree of inventive faculty. This is shown to some extent on the face of his patents, the list of which is as follows: November 11, 1851, "for improvement in stave jointing-machines;" May 22, 1855, "for stave machines;" April 28, 1864, "for improvement in mill-stones;" May 12, 1863, "for improved machine for leveling the faces of mill-stones;" December 12, 1865, "for improvement in nail-plate feeders;" November 20, 1866, "for improvement in faucets;" November 19, 1867, "for improvement in nail-feeding device."

His own testimony, given in an interference proceeding in the patent-office in 1879, shows that none of his inventions were sufficiently meritorious to prosper vigorously. That proceeding involved a question of priority of invention between himself and one Hauck, respecting an improvement in a faucet. He had filed his application for a patent in January, 1879, and undertook to carry back the date of his invention to 1869. The scope and range of his inventive faculty became a subject of inquiry. He there testified that he had made, "he might say, fifty inventions, and had patented over a dozen."

He was cross-examined respecting certain inventions to show that they did not work satisfactorily. He was then asked: "Since 1866, what machines have you conceived and perfected that have worked satisfactorily?" He answered: "To the best of my knowledge, I think they all have. The nail-machine gave satisfaction. I had it running in the works, but the nailers drove it out. The tram and red-staff was a good machine, and adopted by a number of millers. The magnetic clock I consider a good thing, but I am not through with experiments on it yet. I believe this last faucet to be a good thing." If his nail-machine had induced the workmen to drive it out of the shop, it ought to have commended itself to the capitalist. His magnetic clock had not been patented at this time, though it had been for a time the wonder and admiration of the community in which he lived; but when it was patented in 1879 it was as a "new article of manufacture," consisting of a galvanic battery for electrical clocks, which had two old elements united, instead of being disconnected, as in former devices. The history of this clock shows clearly that it was of no practical merit; and the clock had been substantially described in Tomlinson's Encyclopedia; and he had the book before he made his alleged invention. His other electric devices he never patented; and in his testimony in the interference proceedings he did not refer to them as among his perfected and successful inventions. One of these was his magneto-electric machine for short-line telegraphing and fire-alarms, sometimes mentioned as his "magneto key." It was not a new device, and the proofs show that it was a failure.

When the speaking telephone was first introduced to the attention of the scientific public it was pronounced by one of the most eminent electricians of the day "a result of transcendent scientific interest," and "the greatest by far of all the marvels of the electric telegraph." The inventions attributed to Drawbaugh include not only the conception of the principle of the unbroken undulatory electric current, and of the delicate and complex instrumentalities essential to its efficient application in transmitting and reproducing articulate speech, but also of many other devices involving a nice adjustment of forces and requiring sensitive mechanism. These were inventions of a peculiarly scientific order, which would seem to demand a special conversance with the principles of acoustics and electricity. Besides making the cardinal discovery of the theory of the unbroken undulatory current, Drawbaugh is assumed to have perfected a brilliant and extraordinary series of original discoveries, for which, to use the words of Mr. Benjamin, "there is no parallel instance in the whole history of invention." Mr. Benjamin, referring to the microphone, which was introduced to the public in 1878 by Mr. Blake, but which is one of the instruments asserted to have been invented by Drawbaugh at an earlier date, says: "It was looked upon as a great and orginal discovery."

It was said by Chief Justice Taney, (*O'Reilly* v. *Morse*, 15 How. 111,) speaking of the invention of the telegraph:

"No invention can possibly be made consisting of a combination of different elements of power without a thorough knowledge of the property of each, and the mode in which they operate on each other. For no man ever made such an invention without having first obtained this information, unless it was discovered by some fortunate accident."

None of Drawbaugh's alleged discoveries were made by accident. His statement is, that, starting with the belief that speech could be transmitted by electricity, he made first one contrivance and then another, gradually obviating difficulties and making advances experimentally, until he finally perfected the several inventions. In view of Bell's special equipment for investigation and experiment in electrical and acoustic science it would not seem strange that his persistent efforts to effect the electrical transmission of speech were eventually successful, were it not that others as intelligent, as well equipped, as ingenious, and as persevering as he, had devoted years to the same object in vain. He had the assistance of Mr. Watson, an expert in electricity, and a skilled workman in electrical mechanism, in constructing the apparatus employed in his experiments, and who also aided him in his experiments. He had demonstrated his inventive proficiency by inventions in telegraphy for which patents were granted to him. And yet had it not been for an accidental discovery made by him in June, 1875, and which would probably have escaped one whose trained faculties were not centered on a careful study of the phenomena, he might have failed.

Drawbaugh, on the other hand, was not only untutored, but he was isolated by his associations and occupations from contact with men of advanced science; he had narrow opportunities for instruction, and few incentives for profound research. Among the multitude of his inventive conceptions was one that a talking-machine was a possibility. According to the testimony of Lory, a witness for defendants, before Drawbaugh began his practical experiments he exhibited a sketch of a machine that he was about to make that would talk a distance of 20 miles, and work something like a telegraph. If this is true, he commenced on his telephone as the architect plans a building, or the engineer makes a draught of his structure. His own testimony shows that he did not attempt to qualify himself for electrical inventions by any systematic study after he began experimenting with his talking-machine. Although he had undoubtedly acquired considerable desultory information about electricity, and especially about the mode of operation and detail of construction of electrical mechanism, it is obvious that when he commenced with his talking-machine he was a tyro in electrical science, essaying the most difficult work of the electrician. It is almost incredible that the subtle intellectual discoveries which were a closed book to the ablest electrician could have been reached by a smatterer in science, or by any series of empirical experiments. As has been remarked, he seems to have discovered nothing accidentally; yet from the beginning to the end of his narrative there is nothing to indicate the conceptive origin or the mental growth of the alleged invention. He presents a number of devices in the chronological order of their production, and testifies that he made one, and then another and another as experiments led him to modifications and improvements. He cannot describe what receiver or other apparatus he used with his first transmitter, and testifies:

"I had a number of crude apparatuses, but can't remember exactly the shape of any of them. I had membranes stretched over hoops,—over a hoop, I remember that; and I had electro-magnets, and the arrangement was varied. I don't remember exactly the arrangement."

He testifies that when he used the cup-machine he used it in a continuous electric circuit, and thinks he used it as a receiver with Exhibit B as a transmitter. He states that he succeeded in transmitting speech with these two instruments, and, of course, he could only have done this by employing the unbroken undulatory current of electricity. He cannot state how he conceived the initial idea of the undulatory current and the continuous circuit, or, subsequently, the theory of any of the remarkable devices which he produces. His answers to questions intended to elicit such information may be illustrated by giving one of them:

"I don't remember how I came to it. I had been experimenting in that direction. I don't remember of getting at it by accident either. I don't remember of any one telling me of it. I don't suppose any one told me."

He produces sketches or models or originals of instruments which he says he made from time to time. He states that they were used to talk through on various occasions; and from these outlines of accomplished facts leaves the history of his inventions to be filled out by inference and conjecture. An inventor can hardly forget the process of thought by which a great intellectual conception germinates and matures into the consummate achievement; but Drawbaugh's memory is a blank. If the untutored mechanic educated himself into an accomplished electrician by his own experiments and observations, the incidents and phenomena which revealed new discoveries, and illumined the way for new advances, would be indelibly impressed upon his mind. It seems a little short of the miraculous that a man of his capacity and equipment should have produced these inventions at all; more marvelous still that he should have produced them without any intellectual perception of his discoveries.

Another group of important facts which are satisfactorily shown by the proofs are those which indicate Drawbaugh's own knowledge that he was not an original inventor of the telephone. Reference has been made to some of the evidence bearing upon his neglect to patent or *caveat* his invention in discussing the question of his credibility as a witness. If no honest and reasonable explanation can be given for his conduct, the inference is very strong that he knew he did not have a practical telephone to patent. He may have had a talking-machine which was well calculated to excite the curiosity of the community in which he lived; he may have indulged in expectations that in time he could succeed in making a practical speaking telephone, and reap fame and profit from it; but his conduct is almost decisive against the supposition that he had even deluded himself with the belief that he had produced anything sufficiently practical and valuable to patent. He never attempted to exhibit it outside of his own shop to prove that it would transmit speech at a distance of even a quarter of a mile. The proofs show that during all the years from 1867 to 1878 he did not attempt to avail himself of opportunities for demonstrating his invention and bringing it to the notice of friends who were peculiarly qualified to appreciate, and were favorably circumstanced to assist him. One of these persons was Mr. Kiefer, who resided at Harrisburg from 1863 to 1881, and during that period had charge of the telegraphs of the Pennsylvania Railroad Company, and was a member of a firm whose business was the manufacturing of fine electrical machinery. In 1873 he put up a fire-alarm system for that city. Drawbaugh made his acquaintance in 1874 or 1875, and brought his magneto fire-alarm to Mr. Kiefer for examination. At another time he brought the works of his electric clock. He visited Mr. Kiefer on various occasions, obtained small supplies from him, and habitually conversed with him upon the subject of his electrical contrivances. The period of these visits begins just about the time when, according to the theory of the defendants, Drawbaugh had constructed Exhibits E and D, and the invention was complete. He never mentioned to Mr. Kiefer the fact that he had experimented with a telephone. Mr. Wilson was superintendent of telegraphs for the Northern Central Railway Company at Harrisburg from 1864 to 1875. He was also mayor of Harrisburg. The company had an electrical work-shop and supply establishment there for Mr. Wilson's department between 1871 and 1875. During this time Drawbaugh often came to the supply shop and talked with Mr. Wilson about electrical experiments, and obtained parts of batteries, coils, magnets, and other electrical material which the company had cast aside. He brought Mr. Wilson his electric clock and his magneto-electric key, and tried his machine for short-line telegraphing at Mr. Wilson's office. He talked with him frequently about his inventions, but he never mentioned the telephone. His relations with David A. Houck were such that the latter procured him an opportunity to test his magneto key at the telegraph office of the railroad company at Mechanicsburg. Mr. Stees was the superintendent of a car company at Harrisburg, having shops in different parts of the city connected by telegraph lines. He was the first person to employ Bell's telephone on these lines when they were introduced into Harrisburg, late in 1877 or early in 1878. He was a friend of Drawbaugh, and Drawbaugh would naturally have applied to him if he wanted to test his telephone publicly and practically. Isaac Lloyd was a school-teacher and an alderman at Harrisburg; had known Drawbaugh long; was accustomed to visit his shop from time to time; saw many of Drawbaugh's inventions; was present on one occasion when Drawbaugh experimented with his magneto device for telegraphing at Mr. Wilson's telegraph office. Drawbaugh visited him frequently, and they were accustomed to converse about Drawbaugh's inventions. Drawbaugh showed him his dial telegraph, his electric fire-alarm apparatus, and numerous other inventions. Witness assisted him about the electric clock. He was an owner of patents, and a friend to whom Drawbaugh applied for loans, and was interested in mechanical subjects generally. The only mention ever made to him by Drawbaugh about a telephone was in 1878, when Drawbaugh told him he was experimenting with a telephone. From 1867 to July, 1873, Drawbaugh was intimately connected with the persons composing the Drawbaugh Manufacturing Company, which was engaged in manufacturing devices under Drawbaugh's patents. He was a stockholder and the master mechanic of this company. Among the officers and stockholders were many men of capital and enterprise. There came a time when the managers of the company wanted Drawbaugh to suggest new devices for the company to manufacture. He never suggested the telephone, nor attempted to induce the managers of that company to investigate or exhibit his talking-machine. A number of the managers and employes of this concern testify that they never heard of the existence of the talking-machine during the life of the company.

Without attempting to refer to other testimony to the same general effect, what has already been referred to shows that if Drawbaugh had seriously desired to bring his talking-machine into public notice, and secure the fruits of his invention, he had ample opportunity to do so. Who can doubt that if he had a practical telephone to exhibit he would have selected just such men as Kiefer, Wilson, and the others, to demonstrate it to them, and enlist them to demonstrate its utility and value to the public. Such an invention was of a kind well calculated to excite public interest, and to impress practical men with a quick appreciation of its commercial importance and its pecuniary value. It was so sufficiently perfected, according to the theory of the defense, that a patent could have been obtained prior to 1870 to secure the application of the principle, and to compel every subsequent inventor to pay tribute to the discoverer of a new art. For years it was mechanically perfect, and its efficiency and importance as a great factor in human intercourse could have been demonstrated to the public without appreciable inconvenience or expense. Drawbaugh fully appreciated its importance and value. He had the means to patent it himself, and friends to assist him in introducing it into public use. He had the talent to induce others to invest in his inventions. No explanation is possible why, under such circumstances, his efforts should have left no mark upon the annals of inventive progress, and given no evidence of life beyond the idle curiosity his talking-machine excited in the circle of his admirers during all these years. His conduct is more persuasive to show that he did not have a practical, operative telephone, than the testimony of a multitude of witnesses who may have seen and heard talking-machines at his shop during this period. But the complainant has given evidence of his declarations made by him before he had any interest to pervert the truth, which afford a reasonable explanation of his conduct, and go far to explain how the testimony of the corroborative witnesses may be reconciled with the truth.

In 1874–76 Drawbaugh issued a business card advertising himself as "inventor, designer, and solicitor of patents." On the back of this card is printed a list of his inventions as follows: "Stave-heading and shingle cutter; barrel machinery; stave jointing-machine; tram and red-staff for leveling face of mill-stones; rine and driver for running mill-stone; nail machinery for feeding nail-plates; pumps, rotary and others; hydraulic ram; the Drawbaugh rotary measuring faucet; carpet-rag looper; electric clock; and magneto-electric machine for short-line telegraphing and fire-alarm, and propelling electric clocks." He takes pains to say of this magneto-machine on his card that it "can be applied to any form of electric movement, and dispenses with a galvanic battery." He had obtained patents for some of these inventions, but had not for others. He was then experimenting with his electric clock and with his magneto-machine for short-line telegraphing, fire-alarms, etc., and included them in the list of his inventions. The omission to mention the most important one of all of his inventions—one respecting which, according to his present testimony, there had not been a week from the time he made his first cup-machine that he had not been engaged with it—one which was complete before his electric clock was complete—is a significant statement by implication that he had no such invention to advertise. It is to be remembered that when he chronicled his achievements in the autobiographical sketch of 1878 the Bell telephone had been introduced into commercial use at Harrisburg, three miles from Drawbaugh's shop, and the local newspapers had been full of the subject. The cursory allusion in that autobiography to "several kinds of telephones" is in striking contrast with the eulogistic description of the electric clock, and wholly inconsistent with the theory that he deemed himself to be the originator of the telephone which at that particular time was a topic of universal interest.

In his testimony given in 1879, in the interference proceeding with Hauck, although he did not include the talking-machine in the category of his successful inventions, in the course of his testimony he produced a sketch of his faucet, and stated that he made it "about 1874 to 1876, when I was experimenting on telephones or phonographs." He represented himself, not as an inventor of that which he is now claimed to have perfected, but as an experimenter with a "telephone or phonograph." It is instructive to read this statement in juxtaposition with a statement made by him to Mr. Matthews in the preceding year. Mr. Matthews was the managing editor of the Baltimore *American*, and in April, 1878, made a visit to Drawbaugh at his shop to see Drawbaugh's magnetic clock, in consequence of information received from a correspondent. He was a careful observer, who went there obviously for the purpose of writing an article for his paper. That his memory is unusually retentive and accurate, and that he is a careful and conscientious man, is apparent from a letter written by him in December, 1883, after the proofs in the case had been closed, and in which he manifests a desire to correct certain errors of detail in his deposition. Upon that visit his attention was chiefly directed to the clock; but he examined Drawbaugh's other inventions, and conversed with him about them, and, among other things, conversed about the telephone. Drawbaugh's statement to him on that occasion was that he had invented apparatus to send messages by means of an alphabet founded upon difference of sounds. He did not profess to be the inventor of the speaking telephone, or assert that he had ever transmitted speech successfully with his apparatus. He said that the idea of transmitting sounds in this manner was not new, and that he had read of it some years before in a publication translated from the French, and he denied Bell's right to claim the invention of the telephone, because of that publication. In the article founded on that interview, which Mr. Matthews subsequently wrote for publication in the Baltimore *American*, he adverts to the several useful agricultural and mechanical devices patented by Drawbaugh, and adds:

> "It may be mentioned that Mr. Drawbaugh constructed a rude telephone long before Mr. Edison loomed up as the 'boss' inventor. He never expected to send articulate sounds over a magnetized wire, but he believed that an alphabet could be arranged after the manner of the musical scale, and that messages could be transmitted and understood by the variations of the tone and pitch. This unlettered mechanic came very near anticipating Edison and Bell in the invention of the telephone, and nothing but his poverty prevented him from conducting his experiments to a successful issue."

His advertising card, his testimony before the patent-office, his autobiography, and his statement to Matthews, authenticated in writing, were all made when he had no pecuniary interest to color the facts, and upon occasions when he was anxious to present himself in the most favorable light as an inventor; and they were all made after his talking-machine, according to the theory of the defendants, was a perfected invention, and known to be such by many of his friends and neighbors. These are declarations evidenced in writing, and one of them made under oath, which point in but one direction. They are consistent with his conduct. They show that he understood himself to be an experimenter with telephones or phonographs, but not the inventor of the speaking telephone. The complainant has supplemented this evidence by the testimony of other substantial witnesses who had favorable opportunities to know what Drawbaugh had invented, and who describe what they saw and did not see at his shop, and narrate what he said about his talking-machines on various occasions. This testimony indicates that at as late a period as in 1878–79 Drawbaugh was an experimenter, but not the author of the parent invention, nor one who had perfected any valuable improvement upon it, and is in substantial accord with his statement to Mr. Matthews and his testimony in the interference proceedings. What gives point and force to this testimony, and parries the ordinary objections to the reliability of verbal declarations, is that these witnesses are persons who would have been forcibly impressed, because of their interest in the particular subject, by any assertion by Drawbaugh that he was an inventor of the telephone. During the time in question Drawbaugh had friendly relations with the newspapers of the vicinity, his friends were frequently communicating laudatory notices of his mechanical and inventive efforts to the press, and he himself visited one of the newspaper offices in the spring of 1878 to show a telephone he had made. These newspapers had published articles about the Bell telephone, but up to the spring of 1878, while many notices had been published in them about his electric clock and other inventions, describing him as a man of extraordinary genius, there had been no mention of the telephone, and when in the spring of 1878 the subject was mentioned, he was referred to as one who was "inventing a telephone on a different plan from that now occupying the attention of the scientists," and as about completing "the new telephone he is now constructing."

In this connection it is to be noted that soon after telephones were introduced in Harrisburg, late in 1877 or early in 1878, Drawbaugh visited the offices where they were used, examined the inside of the instruments, and borrowed one to take home, which he kept for several days; and the instrument which he borrowed bears a close resemblance in appearance to Exhibit A, which, it is asserted, he had made in 1873 or 1874.

No extended reference will be made to the testimony of other witnesses, such as Mr. Weaver and Mr. Grissinger, showing declarations of Drawbaugh, made after the Bell telephone was in commercial use, to the effect that although he had experimented on the telephone years before Bell he had obtained no satisfactory results. It remains to consider what effect is to be given to the testimony of the multitude of witnesses who have been produced to substantiate the defense. Disregarding the testimony which is merely hearsay, and therefore incompetent as evidence of the main fact, the testimony of many other witnesses is overthrown by the palpable improbability of their statements, or by the contradictions between their statements and those of other witnesses for the defendants upon substantive points, or by successful attacks upon their accuracy in the rebutting testimony of the complaint. There still remains a formidable number of witnesses who testify to seeing or using Drawbaugh's talking-machine, and some of whom identify particular exhibits as the instruments which they saw or tried. No doubt is entertained that Drawbaugh was experimenting at an early period with telephones or phonographs. He knew about the phonograph or phonautograph of Scott as early as in 1863. The membrane diaphragm excited by sonorous waves, and the mechanism of the phonograph were not novelties, and, among the diversity of inventive possibilities, had probably attracted his interest. Prior to the issue of Bell's patent, Dr. Van De Weyde had made public experiments with the Reis telephone at the city of New York, and others had made like experiments elsewhere. In May, 1869, a full description of the instrument and of the experiments was published in the newspaper, *The Manufacturer and Builder*, treating it as a highly interesting curiosity which contained the germ of great practical purposes. Whether other newspapers noticed the experiments or not is not shown, nor is it shown that Drawbaugh saw the article in *The Manufacturer and Builder*. It would be difficult to prove the circumstances if he did see it. Some such publication probably stimulated him to experiment. If he made a sketch of the mechanism at the start the material for it was at hand. As is stated by Mr. Benjamin, it has been asserted of the Reis instrument that certain sounds of the human voice can be transmitted by it; but in truth these are merely fragmentary reproductions of vocal sounds, and the transmission of articulate speech could not be effected because it was constructed on the make and break principle, instead of on that of the undulatory unbroken current.

It is not strange to any reader of the autobiography that Drawbaugh should have taken up the telephone. That he and those about him should have treated it as a talking-machine is entirely natural. That his talking-machine, as late as in 1876, bore a striking resemblance to the Reis telephone is shown by Mr. Shapley's testimony, a witness who noticed the resemblance, and loaned Drawbaugh a copy of the *Scientific American* describing it.

There is enough here to explain Drawbaugh's declarations to his neighbors about the talking-machine he was inventing, and to excite the curiosity of the community. A careful reading of the proofs renders it easy of belief that the witnesses who testify about casual visits to his shop, which occurred many years before their testimony was delivered, and to cursory tests of his instruments on those occasions, have confused the fragmentary and incoherent articulation of such an apparatus, with the hearing of distinct words and sentences. When witnesses undertake—as many of them do—to give the exact words or sentence heard in the instrument five or ten years before, when their attention was not called to the subject afterwards, no hesitation is felt in rejecting such statements as utterly incredible. It may be charitably inferred that such a witness has confused his recollection with more recent impressions. As will hereafter be shown, the proofs demonstrate that most of the witnesses who testify to having heard distinctly and coherently through the talking-machine—all those who indicate the Exhibits B, F, and C as the instruments—are mistaken, if they are truthful. If Drawbaugh was a charlatan, he may have assisted in deluding them; and the proofs show that between 1872 and 1874 a string telephone was in his brother's shop in the village. The fact that he never attempted to exhibit his machine outside of his shop, where it could be used between points some considerable distance apart, and where its real capacity could be readily observed, is significant in this connection.

The more important testimony is that by which it is sought to identify the several exhibits and show their existence at times consistent with the theory of the natural evolution of the invention. The identification of particular exhibits as seen by the witnesses among the various objects of curiosity at Drawbaugh's shop several years before they testify, is necessarily unreliable when it is attempted by observers who had no knowledge of the mode of operation or of the internal organization of the instruments. Such witnesses could not appreciate what they saw, even if they examined the instruments. Most of the witnesses belong to this class. Indeed, the greater proportion of them do not profess to identify the exhibits positively. Some are more certain than others that particular exhibits are the instruments they saw. Exhibits F, B, and C are fragmentary remains of instruments, and their value depends upon Drawbaugh's description of the operative parts that no longer exist. Scores of witnesses testify to seeing the tumbler device resembling Exhibit F, and the tin-can device resembling Exhibit B, but the identification of the other exhibits prior to the date of Bell's patent is comparatively feeble. The appearance of Exhibits F, B, and C is sufficiently peculiar and distinctive to impress the memory of those who saw them. On the other hand, the other exhibits are not of this character, and all that ordinarily the witnesses can safely say of them is that five years or more before testifying they think they saw or used a small walnut box externally resembling I, or A, or E, or D.

It may be said generally of all the testimony of the witnesses who attempt to identify exhibits, that it is mainly valuable when it proceeds from those who used the instruments which they think they remember, and obtained results. They must remember the results obtained much better than the minor differences of appearance presented by the instruments. Granting that Exhibits F, B, and C would be likely to be remembered, what shall be said of the value of the testimony of scores of witnesses who state that they tested these instruments, or saw others test them, and they articulated perfectly, when it appears by the most authentic test that these instruments were incapable of such articulation?

In March, 1882, after most of the proofs in the case had been taken, a test was made of the capacity of the exhibits to transmit speech in the presence of the counsel and the experts for the respective parties. It is not accurate to say a test was made of the exhibits, but reproductions of F, B, and C were made by Drawbaugh, and as rehabilitated by him were used for the test. Whether these were honest reproductions no one can tell; but, such as they were, they were experimented with by Drawbaugh before they were subjected to the test. Whatever else that test demonstrated, it proved that articulate speech could not have been practically communicated through Exhibits F, B, and C at Drawbaugh's shop, under similar conditions, and that only fragmentary or incoherent speech could be occasionally and exceptionally rendered by the reproduced instruments, which had been experimented with privately before the public test. The proofs show that all along to 1878 Drawbaugh exhibited his earlier instruments, F and B, to spectators, and used them as his talking-machine, sometimes showing or using both together, and sometimes one of them. The testimony of the defendants' witnesses, Springer, Moore, Musser, and Bayler, is pertinent upon this point, and has been referred to. How is it to be explained that he used these crude instruments in 1875 and 1876 as his talking-machine, if he had the better instruments, especially such instruments as E and D? But, in view of the fact now shown, that these earlier exhibits are incapable of satisfactory articulation, what confidence can be placed in the rest of the testimony produced to identify exhibits? If the witnesses are mistaken in identifying these very characteristic instruments, and in recalling the results obtained through them, little reliance can be placed upon the identification of other instruments, or upon the statement of the results which the witnesses think were obtained through them. If these witnesses are mistaken in the dates which they fix for the occasions they speak of, their testimony can be reconciled with all the probabilities of the case. And the reasonable explanation of their testimony is that those witnesses who really saw or used the later exhibits did so in 1876, 1877, 1878, and later, instead of on earlier occasions.

The proofs on both sides lead to the general conclusion that Drawbaugh was not an original inventor of the speaking telephone, but had been an experimenter, without obtaining practical results until the introduction of the instruments into Harrisburg. It is very probable that after reading in the *Scientific American*, loaned to him by Mr. Shapley in October, 1876, the article purporting to describe Bell's telephone, but which really describes better the Reis apparatus, he undertook to improve his old devices. At that time, or after he had examined the telephone instruments at Harrisburg and carried one of them home to study, he may have altered the organization of his instrument and made the intermediate exhibits between F and D. If he exhibited them at his shop, and was able to transmit speech through them, this fact will account for the testimony of the witnesses who identify these exhibits, and may be mistaken as to the time they saw them. The real history of his talking-machine is known only to himself, and it will not be profitable to conjecture when he made the advanced instruments which he claims to have made in February, 1875, and the later instruments. It may be that in discrediting his narrative, and rejecting the theory of the facts which rests upon it, the value of the corroborative testimony has been underestimated. However this may be, no doubt is entertained as to the conclusion which should be reached upon the proofs. Succinctly stated most favorably for the defendants the case is this: One hundred witnesses, more or less, testify that on one or more occasions, which took place from five to ten years before, they think they saw this or that device used as a talking-machine. They are ignorant of the principle and of the mechanical construction of the instruments, but they heard speech through them perfectly well, and through one set of instruments as well as the other. This case is met on the part of the complainants by proof that the instruments which most of the witnesses think they saw and heard through were incapable of being heard through in the manner described by them; and further, that the man who knew all about the capacity of his instruments never attempted to use them in a manner which would demonstrate their efficiency and commercial value, but, on the contrary, for ten years after he could have patented them and for five years after they were mechanically perfect, knowing all the time that a fortune awaited the patentee, and with no obstacles in his way, did not move, but calmly saw another obtain a patent, and reap the fame and profit of the invention. Without regard to other features of the case it is sufficient to say that the defense is not established so as to remove a fair doubt of its truth; and such doubt is fatal.

The observation of an eminent commentator may be quoted as apposite to the case:

> "No form of judicial evidence is infallible, however strong in itself; the degree of assurance resulting from it amounts only to an indefinitely high degree of probability; and perhaps as many erroneous judgments have taken place on false or mistaken direct testimony as on presumptive proof." Best, Ev. § 468.

A decree is ordered for complainant.

Three days after the decision, on December 5, Bell attorney Dickerson entered the interlocutory decree in the Clerk's Office of the United States Circuit Court. The master was ordered to prepare and report an account of the gains made by the defendant since June 30, 1880.

While the People's Company case was moving along, the Bell legal army was fighting on other fronts. The Overland

Telephone Company, a New York corporation, proposed to enter the telephone business with instruments and patents of Myron L. Baxter of Illinois. Branch companies were formed for New Jersey and Pennsylvania. In 1884, American Bell flagged the operation in three circuit courts. By stipulation and agreement, these cases were heard as one in the United States Circuit Court at Philadelphia. Certain evidence became a part of the record of the People's Company case and vice versa.

This maneuver provided the opportunity for People's Company to have a new set of Drawbaugh instruments constructed, officially tested, and the testimony as to their performance introduced into the Overland case record.

Fred M. Ott had been appointed special examiner in the Overland case also and was taking testimony at Philadelphia.

Meanwhile, Daniel and his helper, George W. Heiges, the workman who had helped expert Park Benjamin with the "talking machine" tests at the creek workshop in December, 1881, finished the new set of instruments and took them to Philadelphia. A new expert was hired to go with the new instruments: George F. Barker, professor of physics at the University of Pennsylvania. Testified the professor concerning the "A," "B," "C," and "I" machines:

(Q) To what place did you take those instruments that were so placed in your hands?

(A) They were sent first to my house, and afterward to my laboratory at the university.

(Q) Who was present at those tests made by you at your laboratory, and who assisted you in making said tests?

(A) Generally my professional assistant was present and assisted me, and sometimes one or more of my students. On two occasions, Mr. Drawbaugh, Mr. Heiges, and Mr. Smith were present. . . . As to the results obtained he had this to say: The results which I obtained with these instruments were, in general, entirely satisfactory.

Newly-built "F" microphones also were supplied him for testing. Concerning them, he stated, "The results obtained with both the "F" transmitters, when in good adjustment, were in every way satisfactory."

Official tests in the presence of complainant's counsel were made at the Washington Hotel on the afternoon of February 6. It took but three hours to prove that the new machines could transmit articulate speech successfully. Whole newspaper paragraphs went over the wires with but the loss of a few words. Judge Hill, at the transmitter, read from the *Philadelphia Press,* "General Grant is said to be quite ill. He is affected with a trouble of the throat caused by an excessive use of tobacco." He further reported, "Mr. Drawbaugh, as soon as he just heard me say this, threw his tobacco away." An additional revelation informed the listener that Examiner Ott had a toothache.

It will be recalled that, at the New York tests, transmitter "F" posed quite a problem. Although supported by a block of stone, it was continually out of adjustment. During the Philadelphia tests, the rebuilt versions were hand held at an angle of about 45 degrees, an inch or so from the mouth, and required much less adjusting.

With the conclusion of the tests, complainant's counsel and experts disassembled the magic microphones to examine their construction with critical eyes. The powder cup of one was filled with a mixture of bronze and carbon powder, the other, with plain carbon powder. Nothing wrong with that, as Daniel had used such "low conductors" before, but Storrow claimed to have the answer to their outstanding performance. A bit of chicanery, perhaps, foisted upon the court and the innocent expert, but Storrow later put it bluntly as, "simply a piece of fraud."

It developed that an English clergyman, inventor, Henry Hunnings, had made and patented, in 1881, a granular powder transmitter. When operated at a 45-degree angle, it was one of the most powerful known. His patents had been assigned to George Anders and by him to American Bell Telephone. At the Bell Laboratory, a series of experiments and tests had been run during the autumn and winter of 1881-82. One of the technicians familiar with the instruments went to work for the Drawbaugh Telephone Company (franchise holder under the People's Telephone Company). Perhaps that was the answer; Storrow certainly held no doubts that it was.

By agreement with the complainant, the defendant took

new evidence consisting of thirty-eight depositions and reports of the Philadelphia tests, all a part of the Overland case record, before the circuit court for review. In December, 1885, Justice Wallace announced his decision which follows.

American Bell Telephone Co. *v.* People's Telephone Co. and others.[1]

(*Circuit Court, S. D. New York.* 1885.)

1. Patents for Inventions—Evidence—Prior Use.
   Where the evidence of new witnesses (as to a prior use) introduced on rehearing was merely cumulative and less persuasive than the evidence to the same point on the former trial, *held*, that the decree should not be disturbed.
2. Same—Reduction to Practical Use.
   The testimony of witnesses as to the practicability of alleged prior devices incredible where a test of reproductions of those devices showed them to be practically inoperative.
3. Same—Failure to Make Practicable Machine.
   Where the alleged prior inventor was shown to be very near the realization of the invention, but could not, at a period long subsequent, make a practical machine embodying the invention, it was *held*, that his operations were not sufficient to defeat a patent, although witnesses testified that they were successful.
4. Same—Bell Telephone.
   Drawbaugh's talking-machine *held* insufficient as a prior use.

In Equity.

E. N. Dickerson and *E. J. Stower*, for complainants.

Henry C. Andrews and *Lysander Hill*, for defendant.

Wallace, J. Since the decision of this case, in December last, additional proofs have been taken on the part of the defendants, and by consent of complainants have been presented for the further consideration of the court after argument of counsel. All the new evidence is cumulative merely. Such as consists of the testimony of new witnesses to knowledge of the existence of Drawbaugh's talking-machine prior to the date of Bell's patent is far less persuasive than much which has already been considered and rejected as incredible. Such as is intended to show that the earlier instruments made by Drawbaugh were capable of practical use in transmitting articulate speech is more important. The instruments which were introduced as reproductions of lost originals, representing Exhibits B, F, and C, and which were constructed by Drawbaugh, or under his supervision, as accurate reproductions of the originals, were shown to be practically inoperative. After they had been introduced in evidence, and the expert witness for the defendants had testified that he had found them to be good practical instruments, a test of their efficiency was made at the request of counsel for the complainants which took place in the presence of the experts and of the counsel for both parties. This test, which was made in 1882, indicated decisively that the testimony of a large number of witnesses respecting experiments with Drawbaugh's earlier instruments on various occasions in the years 1869 to 1876 was unreliable. While, according to the testimony of these witnesses, the instruments invariably operated satisfactorily

[1] Reported by Charles C. Linthicum, Esq., of the Chicago bar.

upon the occasions of which the witnesses spoke, and no difficulty was experienced or observed by them in the transmission of the messages sent over or heard through the instruments, it was demonstrated by the experiment, in 1882, that under more favorable conditions speech could only be occasionally transmitted by the reproduced instruments, and that, generally, what was spoken, if transmitted at all, was transmitted in a fragmentary or incoherent form. Whether these instruments were, in fact, reproductions of original instruments was a question which depended solely upon the testimony of Drawbaugh. But assuming that his testimony was true, and that they were reproductions, it was impossible to doubt that they were the best that Drawbaugh knew how to construct after he had become a much more accomplished electrician than he was when, according to his testimony, he made the originals.

Since the former decison new reproductions of these alleged originals have been made from the remnants of the originals, supplemented by drawings verified as correct by Drawbaugh from memory, and it now appears that these reproductions are practical instruments, and fully capable of transmitting articulate speech successfully. It does not appear that Drawbaugh made them, or who made them; and no testimony has been introduced to explain the reason of the failure of the instruments made by Drawbaugh to operate satisfactorily. It appears from the testimony of the only witnesses adduced to the point that, upon a comparison of the former and the recent reproductions, the recent ones are improved in finish and detail of mechanism in respects calculated to improve the efficiency of the apparatus; and it also seems probable that the increased efficiency of Exhibit F, the more important of the reproductions, is due to the employment of a new conducting material between the back plate of the cell and the diaphragm of the transmitter, adapted for the adjustment in a mode never practiced by Drawbaugh, and made the subject of a patent to Henry Hunnings of the date of August 30, 1881. The legitimate effect of the evidence is to show that Drawbaugh was very near the realization of the invention if he had really constructed instruments like the Exhibits F, B, and C prior to the date of Bell's patent. It does not, however, alter the fact that he was unable to make such instruments at a period long subsequent to the time when he claims to have made them; and, in view of this fact, the evidence does not tend to materially fortify the testimony of the witnesses who think, or profess to think, that they heard or saw efficient practical instruments in operation at Drawbaugh's shop on the occasions to which they refer.

The conclusions which were reached at the former hearing have not been modified, and the decree ordered should not be disturbed.

On March 1, 1886, at a stated term of circuit court, Justice Wallace handed down his final decree. People's Telephone was held to have infringed the fifth claim of the first Bell patent and the fifth, sixth, and eighth claims of the second. Complainant waived an accounting of defendant's gains, accepted one dollar as damages, received a judgment with its execution against the defendant in the amount of $10,899.30 for costs, and a perpetual injunction.

On July 16, the defendant appealed from the whole decree of the court; the following day Justice Wallace allowed it. Jones and Sinclair, the "down east" sureties on the appeal bond, each swore a worth of $15,000 over and above his just debts and liabilities; His Honor issued the citation of appeal to American Bell and the case was on its way to the Supreme Court with four other circuit court losers. Amos E. Dolbear, along with the Molecular, Clay Commercial and Overland telephone companies, would be heard by the High Court, too.

On January 24, 1887, proceedings before the nation's highest tribunal began. Only seven of the nine justices participated in the telephone cases. Chief Justice Morrison R. Waite, and associate justices Samuel F. Miller, Stephen J. Field, Joseph P. Bradley, John M. Harlan, Stanley Matthews, and Samuel Blatchford were on the bench. Associate justices Woods and Gray did not sit. It was more than a year before the long expected and eagerly awaited decision was reached. When it came, American Bell won by only one vote!

Justices Waite, Matthews, Miller, and Blatchford constituted the majority; Bradley, Field, and Harlan didn't agree.

Justice Woods, who had disqualified himself because of sickness, died May 14, 1887. Chief Justice Waite, who had written the decision, was too ill to read it but attended the March 19 session to hear Justice Blatchford do it for him. Four days later, Morrison Remich Waite was dead.

The important decision in full follows:

DOLBEAR *et al.* *v.* AMERICAN BELL TEL. CO.[1] MOLECULAR TEL. CO. *et al.* *v.* SAME.[2] AMERICAN BELL TEL. CO. *et al.* *v.* MOLECULAR TEL. CO. *et al.*[3] CLAY COMMERCIAL TEL. CO. *et al.* *v.* AMERICAN BELL TEL. CO. *et al.* PEOPLE'S TEL. CO. *et al.* *v.* SAME.[4] OVERLAND TEL. CO. *et al.* *v.* SAME.

(March 19, 1888.)

1. PATENTS FOR INVENTION—PATENTABILITY—IMPROVEMENTS IN TELEGRAPHY—BELL TELEPHONES—LETTERS PATENT NO. 174,465.

Prior to March 7, 1876, the date of letters patent No. 174,465, to Alexander Graham Bell, for "improvements in telegraphy," it had long been believed by those experienced in the art that, if the vibrations of air caused by the voice in speaking could be reproduced at a distance by means of electricity, the speech itself would be reproduced and understood. Bell discovered that these vibrations could be electrically reproduced by gradually changing the intensity of a continuous current of electricity, so as to make it correspond exactly to the changes in the density of the air caused by the human voice. He then devised a way in which these changes of intensity could be made, and speech actually transmitted. *Held*, that discovery in finding this art, and invention in devising the means of making it useful, were both involved, and that the patent was properly issued, under Rev. St. U. S. § 4886, providing that "any person who has invented or discovered any new and useful art * * * may obtain a patent therefor."

2. SAME—FAILURE OF EXPERIMENTAL INSTRUMENT TO TRANSMIT SOUND.

The fact that the particular instrument which Bell had, and which he used in his experiments, did not, under the circumstances in which it was tried, reproduce the spoken words so that they could be clearly understood, does not invalidate this patent; the proof being abundant, and of the most convincing character, that other instruments, carefully constructed, and made exactly in accordance with the specifications, did and will operate sucessfully, though not so perfectly as instruments made upon the principle of the microphone, or the variable resistance method.

3. SAME—CONSTRUCTION OF CLAIM—"VOCAL OR OTHER SOUNDS."

The fifth claim of letters patent No. 174,465, of March 7, 1876, to Alexander Graham Bell, for "improvements in telegraphy," is as follows: "The method of and apparatus for transmitting vocal or other sounds telegraphically, as herein described, by causing electrical undulations, similar in form to the vibrations of the air accompanying the said vocal or other sounds, substantially as set forth." *Held* that, though articulate speech was not mentioned by name in the patent, the sound of articulate speech was one of the "vocal or other sounds" referred to, and that the claim embraced the art of transferring to or impressing upon an undulatory current of electricity the vibrations of air produced by the human voice in articulate speech, in such a way that the speech would be carried to and received by a listener at a distance on the line of the current.

4. SAME—EXTENT OF CLAIM.

Said claim is not for the use of a current of electricity in its natural state as it comes from the battery, but for putting a continuous current in a closed circuit, under such conditions that changes of intensity therein may be produced exactly corresponding to the changes of density in the air caused by the vibrations which accompany vocal or other sounds, and for using that electrical condition, thus created, for sending and receiving articulate speech telegraphically. So construed, the claim is not for the mere use of electricity, as distinct from the particular process with which it is connected in the patent.

5. SAME—PROCESS AND APPARATUS—MODE OF OPERATION.

Bell, after describing clearly and distinctly in his patent his process of transmitting speech telegraphically, pointed out two ways in which this might be done: One, by the "vibration or motion of bodies capable of inductive action, or by the vibration of the conducting wire itself in the neighborhood of such bodies;" and the other, "by alternately increasing and diminishing the resistance of the circuit, or by alternately increasing and diminishing the power of the battery." He then said that he preferred to employ for his purpose "an electro-magnet, * * * having a coil upon one of its legs," and proceeded to describe the construction of the particular apparatus, in which the electro-magnet, or magneto method, was employed.

[1] Affirming 15 Fed. Rep. 448, and 17 Fed. Rep 604.

[2] Affirming 32 Fed. Rep. 214.

[3] Reversing 32 Fed. Rep. 214.

[4] Affirming 22 Fed. Rep. 309, and 25 Fed. Rep. 725. See, also, 31 Fed. Rep. 729, and 27 Fed. Rep. 663.

For a report of the proceedings by the United States to set the Bell patent aside, see 29 Fed. Rep. 17, and 32 Fed. Rep. 591.

HELLO! DRAWBAUGH LOSES.

THE SUPREME COURT DECIDES THE GREAT TELEPHONE CASE.

The Bell Company Favored on all the Questions—Justices Bradley, Fields and Harlan Dissent—Justices Gray and Lamar Absent.

WASHINGTON, March 19.—The supreme court affirmed the validity of the Bell telephone patents to-day. The opinion was read by Judge Blatchford, and occupied considerable moro than an hour in the reading. It was known yesterday that the court upon reassembling this morning would at onco hand down its decision, and although the best sources of information confirmed the impression, which has been gaining

41. The telephone decision was big news for some, but it left a saddened, disheartened Daniel.

21 South Second Street, Harrisburg,

THE HALL SHORT LINE TELEPHONE.

No dangerous electricity. *Total cost less than one-fourth the annual rental of Bell telephone.* Especially applicable for short distance lines between offices and factories, warehouses, stores. mills, hotels, &c., &c., or wherever instant and reliable intercommunication between intermediate points is desired. *Agents wanted in every city.*

THE HALL TELEPHONE CO.,
Broadway and Astor Place, New York.
apr 23-3t

42. To others, it meant nothing. The Hall telephone, a glorified acoustical instrument, kept right on talking along.

Held, that the fifth claim, set out above, being, *First*, for the process, and, *second*, for the apparatus, it was not to be confined to the magneto instrument, and such modes of creating electrical undulations as could be produced by that form of apparatus; but that it included both the magneto and variable resistance methods, and the particular magneto apparatus described, or its equivalent.

6. SAME—ANTICIPATION—REIS' APPARATUS.

In the apparatus made by Reis, of Germany, (who was himself anticipated by Bourseul,) in 1860, with its several modifications, as described in his prospectus of 1865, and his catalogue of 1866, and also by Legat, Van der Weyde, Ferguson, and others, an intermittent or pulsatory current of electricity was employed; the transmitter, when actuated by the sound waves, making and breaking the circuit at each vibration. *Held*, that the apparatus was, from its very nature, unable to send and receive articulate speech, and that it was not an anticipation of letters patent No. 174,465 of March 7, 1876, to Alexander Graham Bell for "improvements in telegraphy," the essential elements of which are the employment of the undulatory, as contradistinguished from the pulsatory, current of electricity, to transmit and copy air vibrations corresponding exactly in amplitude, rate, and form to those produced by the human voice, and the apparatus therefor.

7. SAME—MCDONOUGH "CIRCUIT BREAKER."

Nor was the Bell patent anticipated by James W. McDonough; his application of April 10, 1876, for a patent, which was rightly refused, on the ground of anticipation by Reis, making a "circuit-breaker," so adjusted as to "break the connection by the vibrations of the membrane," one of the elements of his inventions.

8. SAME—VARLEY'S ENGLISH PATENTS.

Nor is anticipation to be found in the patents of Cromwell Fleetwood Varley, of London, England,—granted, one, June 2, 1868, and the other, October 8, 1870,—for "improvements in electric telegraphs;" the specifications thereto not indicating that the patentee had in mind "undulations," resulting "from gradual changes of intensity exactly analogous to the changes in the density of air occasioned by simple pendulous vibrations," which was Bell's discovery, and on which his art rests; and it being apparent that Varley's only purpose was to superpose upon the ordinary signal current another which, by the action of the make and break principle of the telegraph, would do the work he wanted.

9. SAME—DRAWBAUGH'S EXPERIMENTS—ABANDONMENT.

In support of the claim of Daniel Drawbaugh that he "was the original and first inventor and discoverer of the art of communicating articulate speech between distant places by voltaic and magneto electricity," it was in evidence that he had never told any one how earlier instruments with which he experimented were made, or what his process was, until he was called as a witness in December, 1881, in the case of *American Bell Tel. Co.* v. *People's Tel. Co.* This was nearly twenty years after he had begun his experiments, nearly seven after he had made and used "perfectly adjusted and finished magneto instruments," and more than five after microphones as good, or nearly as good, as those of Blake, which were not invented until 1878, had been constructed, and kept in his shop. It was also nearly six years after the date of Bell's patent, (March 7, 1876,) more than five after the success of Bell's discovery had been proclaimed at the Centennial Exposition in Philadelphia, four after his process had got into public use, three after it had become an established success, and two after Bell had brought his first suit for infringement. In the mean time, the discovery of Bell had been heralded to the world, and Drawbaugh had had abundant means and ample opportunities to make his claim known. During part of this time he had treated his discovery as of secondary importance, and had devoted himself to the advancement of certain other inventions of his of comparatively small merit. In addition, the instruments of Drawbaugh were fairly tested in March, 1882, and failed to produce satisfactory results, and, when offered in evidence, they were in no condition, being mere "remains." *Held* an abandonment. FIELD, BRADLEY, and HARLAN, JJ., dissenting.

10. SAME—FRAUD IN OBTAINING PATENT.

The charge that after Bell swore to his application on January 20, 1876, and after the application, thus sworn to, had been formally filed in the patent-office on February 14, 1876, an examiner, who got knowledge of the Gray *caveat*, put in afterwards on the same day, disclosed its contents to Bell's attorneys, and that they were then allowed to withdraw the application, change it so as to include Gray's variable resistance method over Bell's signature, and over the jurat, and then to restore the application to the files, thus materially altered, as if it were the original, and all this between February 14th and February 19th, examined, and *held* not sustained.

11. SAME—LETTERS No. 186,787—ANTICIPATION—HUGHES' PRINTING TELEGRAPH.

Letters patent No. 186,787, of January 30, 1877, to Alexander Graham Bell, for "improvements in electric telephony," are for the mechanical structure of an electric telephone, to be used to produce the electrical action on which letters patent No. 174,465, of March 7, 1876, to said Bell, for "improvements in electric telegraphy," rests. The fifth claim of the patent of 1877 is as follows: "The formation, in an electric telephone, such as herein shown and described, of a magnet, with a coil upon the end or ends of the magnet nearest the plate." *Held*, the claim as a whole, being for an electric telephone, in the construction of which the plate or diaphragm, the permanent magnet, sounding box, speaking tube, etc., or any of them, are used, and not for the several things in and of themselves, it was not anticipated by the magnet in Hughes' printing telegraph, as described in Schellen's work.

12. CORPORATIONS—CORPORATE EXISTENCE—EVIDENCE.
The authority granted by act Mass. March 19, 1880, (Laws & Resolves, 1880-81, c. 117, p. 74,) to the persons named therein, to organize as a corporation under the provisions of Laws & Resolves, 1870, c. 224, relating to corporations, gave them the right to select a corporate name; and such persons having selected the name of the American Bell Telephone Co., and brought suit in that name, proof of the special act under which they were incorporated, and a certificate of the secretary of the commonwealth in the form required by Laws & Resolves, 1870, c. 224, § 11, is conclusive evidence of the corporate existence.

Appeal from the Circuit Court of the United States for the District of Massachusetts.

Appeal from the Circuit Court of the United States for the Eastern District of Pennsylvania.

Appeals from the Circuit Court of the United States for the Southern District of New York.

These cases were all bills for infringement of letters patent No. 174,465 of March 7, 1876, to Alexander Graham Bell, for "improvements in telegraphy," and of letters patent No. 186,787, of January 30, 1877, to the same, for "improvements in electric telephony." The decree in each and every case, except that of the Molecular Telephone Company, was in favor of the Bell telephone patents, and in that case the decision (32 Fed. Rep. 214) was adverse only upon the fifth claim of the patent of 1877. The appeal of the American Bell Telephone Company was from that part of the decree. The remaining appeals, five in number, were all taken by the several defendants.

Causten Browne and *J. E. Maynadier*, for Dolbear *et al.* *G. P. Lowrey*, *Wheeler H. Peckham*, and *H. D. Donnelly*, for Molecular Tel. Co. *et al.* *W. W. Kerr*, for Clay Commercial Tel. Co. *et al.* *Lysander Hill*, *Don M. Dickinson*, and *T. S. E. Dixon*, for People's Tel. Co. *et al.* *E. N. Dickerson*, *J. J. Storrow*, and *Chauncey Smith*, for American Bell Tel. Co. *et al.*

WAITE, C. J. The important question which meets us at the outset in each of these cases is as to the scope of the fifth claim of the patent of March 7, 1876, which is as follows: "The method of and apparatus for transmitting vocal or other sounds telegraphically, as herein described, by causing electrical undulations, similar in form to the vibrations of the air accompanying the said vocal or other sounds, substantially as set forth." It is contended that this embraces the art of transferring to or impressing upon a current of electricity the vibrations of air produced by the human voice in articulate speech, in a way that the speech will be carried to and received by a listener at a distance on the line of the current. Articulate speech is not mentioned by name in the patent. The invention, as described, "consists in the employment of a vibratory or undulatory current of electricity, in contradistinction to a merely intermittent or pulsatory current, and of a method of and apparatus for producing electrical undulations upon the line wire." A "pulsatory current" is described as one "caused by sudden or instantaneous changes of intensity;" and an "electrical undulation," as the result of "gradual changes of intensity exactly analogous to the changes in the density of air occasioned by simple pendulous vibrations." Among the uses to which this art may be put is said to be the "telegraphic transmission of noises or sounds of any kind;" and it is also said that the undulatory current, when created in the way pointed out, will produce through the receiver at the receiving end of the line "a similar sound to that uttered into" the transmitter at the transmitting end. One of the means of imparting the necessary vibrations, through the transmitter, to produce the undulations, may be the human voice. Articulate speech is certainly included in this description, for it is an "uttered" "sound," produced by the "human voice." It is contended, however, that "vocal sounds" and "articulate speech" are not convertible terms, either in acoustics or in telegraphy. It is unnecesary to determine whether this is so or not. Articulate speech necessarily implies a sound produced by the human voice; and as the patent on its face is for the art of changing the intensity of a continuous current of electricity by the undulations of the air caused by sonorous vibrations, and speech can only be communicated by such vibrations, the transmission of speech in this way must be included in the art. The question is not whether "vocal sounds" and "articulate speech" are used synonymously as scientific terms, but whether the sound of articulate speech is one of the "vocal or other sounds" referred to in this claim of the patent. We have no hesitation in saying that it is, and that, if the patent can be sustained to the full extent of what is now contended for, it gives to Bell, and those who claim under him, the exclusive use of his art for that purpose until the expiration of the statutory term of his patented rights.

In this art—or, what is the same thing under the patent law, this process, this way, of transmitting speech—electricity, one of the forces of nature, is employed; but electricity, left to itself, will not do what is wanted. The art consists in so controlling the force as to make it accomplish the purpose. It had long been believed that, if the vibrations of air caused by the voice in speaking could be reproduced at a distance by means of electricity, the speech itself would be reproduced and understood. How to do it was the question. Bell discovered that it could be done by gradually changing the intensity of a continuous electric current, so as to make it correspond exactly to the changes in the density of the air caused by the sound of the voice. This was his art. He then devised a way in which these changes of intensity could be made, and speech actually transmitted. Thus his art was put in a condition for practical use. In doing this, both discovery and invention, in the popular sense of those terms, were involved; discovery in finding the art, and invention in devising the means of making it useful. For such discoveries and such inventions the law has given the discoverer and inventor the right to a patent, as discoverer, for the useful art, process, method of doing a thing, he has found; and, as inventor, for the means he has devised to make his discovery one of actual value. Other inventors may compete with him for the ways of giving effect to the discovery, but the new art he has found will belong to him, and those claiming under him, during the life of his patent. If another discovers a different art or method of doing the same thing, reduces it to practical use, and gets a patent for his discovery, the new discovery will be the property of the new discoverer; and thereafter the two will be permitted to operate each in his own way, without interference by the other. The only question between them will be whether the second discovery is in fact different from the first. The patent for the art does not necessarily involve a patent for the particular means employed for using it. Indeed, the mention of any means, in the specification or descriptive portion of the patent, is only necessary to show that the art can be used; for it is only useful arts—arts which may be used to advantage—that can be made the subject of a patent. The language of the statue is that "any person who has invented or discovered any new and useful art, machine, manufacture, or composition of matter" may obtain a patent therefor. Rev. St. § 4886. Thus, an art—a process—which is useful, is as much the subject of a patent as a machine, manufacture, or composition of matter. Of this there can be no doubt, and it is abundantly supported by authority. *Corning* v. *Burden*, 15 How. 252, 267; *Cochrane* v. *Deener*, 94 U. S. 780, 787, 788; *Tilghman* v. *Proctor*, 102 U. S. 707, 722, 724, 725; *Fermentation Co.* v. *Maus*, 122 U. S. 413, 427, 428, 7 Sup. Ct. Rep. 304. What Bell claims is the art of creating changes of intensity in a continuous current of electricity, exactly corresponding to the changes of density in the air caused by the vibrations which accompany vocal or other sounds, and of using that electrical condition, thus created, for sending and receiving articulate speech telegraphically. For that, among other things, his patent of 1876 was, in our opinion, issued; and the point to be decided is whether, as such a patent, it can be sustained.

In *O'Reilly* v. *Morse*, 15 How. 62, it was decided that a claim in broad terms (page 86) for the use of the motive power of the electric or galvanic current called "electro-magnetism, however developed, for making or printing intelligible characters, letters, or signs, at any distances," although "a new application of that power" first made by Morse, was void, because (page 120) it was a claim "for a patent for an effect produced by the use of electro-magnetism, distinct from the process or machinery necessary to produce it;" but a claim (page 85) for "making use of the motive power of magnetism, when developed by the action of such current or currents, substantially as set forth in the foregoing description, * * * as means of operating or giving motion to machinery, which may be used to imprint signals upon paper or other suitable material, or to produce sounds in any desired manner, for the purpose of telegraphic communication at any distances," was sustained. The effect of that decision was, therefore, that the use of magnetism as a motive power, without regard to the particular process with which it was connected in the patent, could not be claimed, but that its use in that connection could.

In the present case the claim is not for the use of a current of electricity in its natural state as it comes from the battery, but for putting a continuous current, in a closed circuit, into a certain specified condition, suited to the transmission of vocal and other sounds, and using it in that condition for that purpose. So far as at present known, without this peculiar change in its condition it will not serve as a medium for the transmission of speech, but with the change it will. Bell was the first to discover this fact, and how to put such a current in such a condition; and what he claims is its use in that condition for that purpose, just as Morse claimed his current in his condition for his purpose. We see nothing in Morse's case to defeat Bell's claim; on the contrary, it is in all respects sustained by that authority. It may be that electricity cannot be used at all for the transmission of speech, except in the way Bell has discovered, and that therefore, practically, his patent gives him its exclusive use for that purpose; but that does not make his claim one for the use of electricity distinct from the particular process with which it is connected in his patent. It will, if true, show more clearly the great importance of his discovery, but it will not invalidate his patent.

But it is insisted that the claim cannot be sustained because, when the patent was issued, Bell had not in fact completed his discovery. While it is conceded that he was acting on the right principle, and had adopted the true theory, it is claimed that the discovery lacked that practical development which was necessary to make it patentable. In the language of counsel, "there was still work to be done, and work calling for the exercise of the utmost ingenuity, and calling for the very highest degree of practical invention." It is quite true that when Bell applied for his patent he had never actually transmitted telegraphically spoken words so that they could be distinctly heard and understood at the receiving end of his line; but in his specification he did describe accurately, and with admirable clearness, his process,—that is to say, the exact electrical condition that must be created to accomplish his purpose,—and he also described, with sufficient precision to enable one of ordinary skill in such matters to make it, a form of apparatus which, if used in the way pointed out, would produce the required effect, receive the words, and carry them to and deliver them at the appointed place. The particular instrument which he had, and which he used in his experiments, did not, under the circumstances in which it was tried, reproduce the words spoken so that they could be clearly understood; but the proof is abundant, and of the most convincing character, that other instruments, carefully constructed, and made exactly in accordance with the specification, without any additions whatever, have operated, and will operate, successfully. A good mechanic, of proper skill in matters of the kind, can take the patent, and, by following the specification strictly, can, without more, construct an apparatus which, when used in the way pointed out, will do all that it is claimed the method or process will do. Some witnesses have testified that they were unable to do it. This shows that they, with the particular apparatus they had, and the skill they employed in its use, were not successful; not that others, with another apparatus, perhaps more carefully constructed, or more skillfully applied, would necessarily fail. As was said in *Loom Co.* v. *Higgins*, 105 U. S. 580, 586, "when the question is whether a thing can be done or not, it is always easy to find persons ready to show how not to do it." If one succeeds, that is enough, no matter how many others fail. The opposite results will show that in the one case the apparatus used was properly made, carefully adjusted, with a knowledge of what was required, and skillfully used, and that in the others it was not. The law does not require that a discoverer or inventor, in order to get a patent for a process, must have succeeded in bringing his art to the highest degree of perfection; it is enough if he describes his method with sufficient clearness and precision to enable those skilled in the matter to understand what the process is, and if he points out some practicable way of putting it into operation. This Bell did. He described clearly and distinctly his process of transmitting speech telegraphically, by creating changes in the intensity of a continuous current or flow of electricity, in a closed circuit, exactly analogous to the changes of density in air occasioned by the undulatory motion given to it by the human voice in speaking. He then pointed out two ways in which this might be done: One, by the "vibration or motion of bodies capable of inductive action, or by the vibration of the conducting wire itself in the neighborhood of such bodies;" and the other, "by alternately increasing and diminishing the resistance of the circuit, or by alternately increasing and diminishing the power of the battery." He then said he preferred to employ for his purpose "an electro-magnet, * * * having a coil upon only one of its legs," and he described the construction of the particular apparatus shown in the patent as Fig. 7, in which the electro-magnet, or magneto method, was employed. This was the apparatus which he himself used without entirely satisfactory results, but which Prof. Cross, Mr. Watson, Dr. Blake, Prof. Pope, and others testify has done, and will do, what was claimed for it, and transmit speech successfully, but not so well, indeed, as another constructed upon the principle of the microphone or the variable resistance method.

An effort was made, in argument, to confine the patent to the magneto instrument, and such modes of creating electrical undulations as could be produced by that form of apparatus; the position being that such an apparatus necessarily implied "a closed circuit, incapable of being opened, and a continuous current, incapable of being intermittent." But this argument ignores the fact that the claim is—*First*, for the process; and, *second*, for the apparatus. It is to be read (1) as a claim for "the method of transmitting vocal or other sounds telegraphically, as herein described, by causing electrical undulations similar in form to the vibrations of the air accompanying the said vocal or other sounds, substantially as set forth;" and (2) as for "the apparatus for transmitting vocal or other sounds telegraphically, as herein described, by causing electrical undulations, * * * substantially as set forth." The method, "as herein described," is to cause gradual changes in the intensity of the electric current used as the medium of transmission, which shall be exactly analogous to the changes in the density of the air occasioned by the peculiarities in the shapes of the undulations produced in speech, in the manner "substantially as set forth;" that is to say, "by the vibration or motion of bodies capable of inductive action, or by the vibration of the conducting wire itself in the neighborhood of such bodies," which is the magneto method; or "by alternately increasing and diminishing the resistance of the circuit, or by alternately increasing and diminishing the power of the battery;" which is the variable resistance method. This is the process which has been patented, and it may be operated in either of the ways set forth. The current must be kept closed, to be used successfully; but this does not necessarily imply that it must be so produced, or so operated upon, as to be incapable of being opened. If opened, it will fail to act for the time being, and the process will be interrupted; but there is nothing in the patent which requires it to be operated by instruments which are incapable of making the break. The apparatus, "as herein described," which is included in the claim, is undoubtedly one in which an electro-magnet is employed, and constructed "substantially as set forth" in the specification. One acting on the variable resistance mode is not described, further than to say that the vibration of the conducting wire in mercury, or other liquid, included in the circuit, occasions undulations in the current, and no other special directions are given as to the manner in which it must be constructed. The patent is both for the magneto and variable resistance methods, and for the particular magneto apparatus which is described, or its equivalent. There is no patent for any variable resistance apparatus. It is undoubtedly true that when Bell got his patent he thought the magneto method was the best. Indeed, he said, in express terms, he preferred it; but that does not exclude the use of the other, if it turns out to be the most desirable way of using the process under any circumstances. Both forms of apparatus operate on a closed circuit by gradual changes of intensity, and not by alternately making and breaking the circuit, or by sudden and instantaneous changes, and they each require to be so adjusted as to prevent interruptions. If they break, it is a fault, and the process stops until the connection is restored.

It is again said that the claim, if given this broad construction, is virtually "a claim for speech transmission by transmitting it; or, in other words, for all such doing of a thing as is provable by doing it." It is true that Bell transmits speech by transmitting it, and that long before he did so it was believed by scientists that it could be done by means of electricity, if the requisite electrical effect could be produced. Precisely how that subtle force operates under Bell's treatment, or what form it takes, no one can tell. All we know is that he found out that, by changing the intensity of a continuous current so as to make it correspond exactly with the changes in the density of air caused by sonorous vibrations, vocal and other sounds could be transmitted and heard at a distance. This was the thing to be done, and Bell discovered the way of doing it. He uses electricity as a medium for that purpose, just as air is used within speaking distance. In effect, he prolongs the air vibrations by the use of electricity. No one before him had found out how to use electricity with the same effect. To use it with success, it must be put in a certain condition. What that condition was he was the first to discover, and with his discovery he astonished the scientific world. Prof. Henry, one of the most eminent scientists of the present century, spoke of it as "the greatest marvel hitherto achieved by the telegraph." The thing done by Bell was "transmitting audible speech through long telegraphic lines;" and Sir William Thomson, on returning to his home in England, in August or September, 1876, after seeing at the Centennial Exposition in Philadelphia what Bell had done and could do by his process, spoke in this way of it to his countrymen: "Who can but admire the hardihood of invention which devised such very slight means to realize the mathematical conception that, if electricity is to convey all the delicacies of quality which distinguish articulate speech, the strength of its current must vary continuously, as nearly as may be, in simple proportion to the velocity of a particle of air engaged in constituting the sounds." Surely, a patent for such a discovery is not to be confined to the mere means he improvised to prove the reality of his conception.

We come now to consider the alleged anticipation of Philipp Reis. And here it is to be always kept in mind that the question is not whether the apparatus devised by Reis to give effect to his theory can be made, with our present knowledge, to transmit speech, but whether Reis had in his time found out the way of using it successfully for that purpose; not as to the character of the apparatus, but as to the mode of treating the current of electricity on which the apparatus is to act, so as to make that current a medium for receiving the vibrations of air created by the human voice in articulate speech at one place, and, in effect, delivering them at the ear of a listener in another place. Bell's patent is not alone for the particular apparatus he describes, but for the process that apparatus was designed to bring into use. His patent would be quite as good if he had actually used Reis' apparatus in developing the process for which it was granted. That Reis knew what had to be done in order to transmit speech by electricity is very apparent, for in his first paper he said: "As soon as it is possible to produce, anywhere and in any manner, vibrations whose curves shall be the same as those of any given tone, or combination of tones, we shall receive the same impression as that tone, or combination of tones, would have produced on us." Bourseul also knew it before Reis, for, in a communication published in a Paris journal in 1854, he said: "Reproduce precisely these vibrations," to-wit, the vibrations made by the human voice in uttering syllables, "and you will reproduce precisely these syllables." Reis discovered how to reproduce musical tones, but he did no more. He could sing through his apparatus, but he could not talk. From the beginning to the end he has conceded this. In his first paper he said: "Hitherto it has not been possible to reproduce the tones of human speech with a distinctness sufficient for every one. The consonants are for the most part reproduced pretty distinctly, but the vowels, as yet, not in an equal degree. The cause of this I will attempt to explain. According to the experiments of Willis, Helmholtz, and others, vowel tones can be produced artificially, if the vibrations of one body are from time to time augmented by those of another, something as follows: An elastic spring is set in vibration by the blow of a tooth on a toothed wheel: the first vibration is the greatest, and each subsequent one is smaller than the preceding. If, after a few vibrations of this kind, (the spring not coming to a rest in the mean time,) the tooth wheel imparts a new stroke, the following vibration will be again a maximum, and so on. The pitch of the tone produced in this way depends upon the number of vibrations in a given time; but the character of the tone, upon the number of swellings in the same time. * * * Our organs of speech probably produce the vowels in the same manner, through the combined action of the upper and lower vocal cords, or of these latter and the cavity of the mouth. My apparatus reproduces the number of vibrations, but with an intensity much less than that of the original ones; though, as I have reason to believe, to a certain degree proportional among themselves. But in the case of these generally small variations, the difference between large and small vibrations is more difficult to perceive than in the case of the original waves, and the vowel is therefore more or less indistinct." And again: "I have succeeded in constructing an apparatus with which I am enabled to reproduce the tones of various instruments, and even to a certain extent the human voice." No one of the many writers whose papers are found in the records claim more than this for Reis or his discoveries. Although his first paper was published in 1861, and Bell did not appear as a worker in the same field of scientific research until nearly 15 years afterwards, no advance had been made, by the use of what he had contrived or of his method, towards the great end to be accomplished. He caused his instruments to be put on the market for sale, and both he, and those whom he employed for that purpose, took occasion to call attention to them by prospectus, catalogue, and otherwise, and to describe what they were, and what they would do. In his own prospectus, which was published in 1865 and attached to the apparatus, he says: "Every apparatus consists * * * of two parts,—the telephone proper and the receiver. * * * These two parts are placed at such a distance from each other that singing or toning of a musical instrument can be heard in no other way from one station to the other except through the apparatus." And, "Besides the human voice, there can be reproduced (according to my experience) just as well the tones of good organ-pipes from F—c, and those of the piano." Albert, the mechanician employed to make the instruments, in his catalogue published in 1866, enumerates, among the things he has for sale, "Telephone of Reis for reproduction of tones by electricity." In a work on electricity by Robert M. Ferguson, published by William & Robert Chambers, London and Edinburgh, in 1867, it is said, in speaking of the telephone: "This is an instrument for telegraphing notes of the same pitch. Any noise producing a single vibration of the air, when repeated regularly a certain number of times in the second, (not less than thirty-two,) produces, as is well known, a musical sound. * * * A person when singing any note causes the air to vibrate so many times per second, the number varying with the pitch of the note he sings; the higher the note, the greater being the number of vibrations. If we, then, by any means, can get these vibrations to break a closed circuit, * * * the note sung at one station can be reproduced, at least so far as pitch is concerned, at another. Reis' telephone (invented 1861) accomplishes this in the following way," which is then described. But it is needless to quote further from the evidence on this branch of the case. It is not contended that Reis had ever succeeded in actually transmitting speech, but only that his instrument was capable of it if he had known how. He did not know how, and all his experiments in that direction were failures. With the help of Bell's later discoveries in 1875 we now know why he failed.

As early as 1854, Bourseul, in his communication which has already been referred to, had said, substantially, that, if the vibrations of air produced by the human voice in articulate speech could be reproduced by means of electricity at a distance, the speech itself would be reproduced and heard there. As a means of stimulating inquiry to that end, he called attention to the principle on which the electric telegraph was based, and suggested an application of that principle to such a purpose. He said: "The electric telegraph is based on the following principle: An electric current, passing through a metallic wire, circulates through a coil around a piece of soft iron, which it converts into a magnet. The moment the current stops, the piece of iron ceases to be a magnet. This magnet, which takes the name of 'Electro-Magnet,' can thus in turn attract, and then release, a movable plate, which, by its to and fro movement, produces the conventional signals employed in telegraphy." Then, after referring to the mode in which speech is transmitted by the vibrations of the air, he said: "Suppose that a man speaks near a movable disk, sufficiently flexible to lose none of the vibrations of the voice; that this disk alternately makes and breaks the connection with a battery,—you may have at a distance another disk which will simultaneously execute the same vibrations." That Reis was working all the time, from the beginning to the end of his experiments, upon the principle of the telegraph as thus suggested by Bourseul, is abundantly proven. Thus, in his first paper, after describing his cubical block apparatus, he says: "If, now, tones, or combinations of tones, are produced in the neighborhood of the block, so that sufficiently powerful waves enter the opening, *a*, then these sounds cause the membrane, *b*, to vibrate. At the first condensation, the hammer-like wire, *d*, is pushed back; at the rarefaction, it cannot follow the retreating membrane, and the current traversing the strips remains broken until the membrane, forced by a new condensation, again presses the strip * * * against *d*. In this way each sound-wave causes a breaking and closing of the current. At each closing of the circuit, the atoms of the iron wire inside the distant spiral are moved away from each other; on breaking the circuit, these atoms seek to regain their position of equilibrium. When this happens, in consequence of the reciprocal actions of elasticity and inertia, a number of vibrations are produced, and they give the longitudinal sound of the rod. This is the case if the making and breaking of the current occur with comparative slowness. If they occur more rapidly than the oscillations of the iron core, due to its elasticity, the atoms cannot complete their course. The paths described become shorter in proportion as the interruptions are more frequent, but then are just as numerous as these. The iron wire no longer gives its longitudinal normal tone, but a tone whose pitch corresponds to the number of interruptions in a given time. This is the same as saying that the rod reproduces the tone impressed upon the interrupter."

Such was the beginning, and it was maintained persistently to the end, as well by Reis as by those who availed themselves of what he was doing. To this the Reis-Legat apparatus forms no exception, for in the paper describing it Legat says: "The operation of the apparatus described is as follows: When at rest, the galvanic circuit is closed. When the air which is in the tube, *a*, *b*, of the apparatus is alternately condensed and rarefied by speaking into it, (or by singing or introducing the tones of an instrument,) a movement of the membrane closing the smaller opening of the tube is produced, corresponding to such condensation or rarefaction. The lever, *c*, *d*, follows the movements of the membrane, and opens and closes the galvanic circuit at *d*, *g*, so that at each condensation of the air in the tube the circuit is opened, and at each rarefaction the circuit is closed. In consequence of this operation, the electro-magnet of the apparatus, in accordance with the condensations and rarefactions of the column of air in the tube, * * * is correspondingly demagnetized and magnetized, and the armature of the magnet is set into vibrations like those of the membrane in the transmitting apparatus." We have not had our attention called to a single item of evidence which tends in any way to show that Reis, or any one who wrote about him, had it in his mind that anything else than the intermittent current caused by the opening and closing of the circuit could be used to do what was wanted. No one seems to have thought that there could be another way. All recognized the fact that the "minor differences in the original vibrations" had not been satisfactorily reproduced, but they attributed it to the imperfect mechanism of the apparatus used, rather than to any fault in the principle on which the operation was made to depend. It was left for Bell to discover that the failure was due, not to workmanship, but to the principle which was adopted as the basis of what had to be done. He found that what he called the "intermittent current"—one caused by alternately opening and closing the circuit—could not be made, under any circumstances, to reproduce the delicate forms of the air vibrations caused by the human voice in articulate speech, but that the true way was to operate on an unbroken current by increasing and diminishing its intensity. This he called a "vibratory or undulatory current," not because the current was supposed to actually take that form, but because it expressed with sufficient accuracy his idea of a current which was subjected to gradual changes of intensity, exactly analogous to the changes of density in the air occasioned by its vibrations. Such was his discovery, and it was new. Reis never thought of it, and he failed to transmit speech telegraphically; Bell did, and he succeeded. Under such circumstances, it is impossible to hold that what Reis did was an anticipation of the discovery of Bell. To follow Reis is to fail, but to follow Bell is to succeed. The difference between the two is just the difference between failure and success. If Reis had kept on, he might have found out the way to succeed; but he stopped, and failed. Bell took up his work, and carried it on to a successful result.

As to what is shown to have been written and done by Dr. Van der Weyde, it is only necessary to say that he copied Reis; and it was not until after Bell's success that he found out how to use a Reis instrument so as to make it transmit speech. Bell taught him what to do to accomplish that purpose.

So, as to James W. McDonough. We presume that it will not be claimed that he is entitled to more than he asked for in his application for a patent, filed April 10, 1876; and there a "circuit breaker," so adjusted as to "break the connection by the vibrations of the membrane," is made one of the elements of his invention. The patent office was clearly right in holding that he had been anticipated by Reis.

The patents of Cromwell Fleetwood Varley, of London, England,—granted, one, June 2, 1868, and the other, October 8, 1870,—were for "improvements in electric telegraphs." The objects of the invention covered by the first were "to cut off the disturbance arising from earth currents, to obtain a high speed of signaling through long circuits, and, should the conductor become partially exposed, to preserve it from being eaten away by electrolytic action;" and the object of the second was the "increase of the transmitting power of telegraph circuits, by enabling more than one operator to signal independent messages at the same time, upon one and the same wire, to and from independent stations." While this patentee in his specification says, "By my invention, I superpose upon the currents used for working the ordinary telegraphs rapid undulations or waves, which do not practically alter the mechanical or chemical power of the ordinary signal currents," and that "these undulations are made to produce distinct and independent audible or other signals so long as these undulations are produced, whether ordinary signal currents be flowing or not," it is apparent that he uses the terms "undulations" and "waves" in an entirely different sense from Bell, for his patent implies operation on the principle of the electric telegraph; that is to say, by making and breaking the circuit. A Morse key, or something equivalent, is to be used; and besides, in the descriptive portion of the patent, it is said: "When the current is flowing through the coils of the electro-magnet, the horns of the fork, *k*, are drawn apart, and the spring, *l′*, loses its contact; then, as the attraction of the magnet ceases, the horns of the fork spring back. This remakes the contact, and so a continual tremor is communicated to the tuning-fork." In short, there is nothing in any part of the specification to indicate that the patentee had in his mind "undulations" resulting "from gradual changes of intensity exactly analogous to the changes in the density of air occasioned by simple pendulous vibrations," which was Bell's discovery, and on which his art rests. Varley's purpose was to superpose, that is to say, place, upon the ordinary signal current, another which, by the action of the make and break principle of the telegraph, would do the work he wanted.

Another alleged anticipation is that of Daniel Drawbaugh. Bell got his patent March 7, 1876, and the fortunate accident which led to his discovery occurred June 2, 1875. Active litigation to enforce his patented rights was begun by his company on the 12th of September, 1878, with a suit, in the circuit court of the United States for the district of Massachusetts, against Richard A. Dowd. This suit was defended by the Western Union Telegraph Company, and vigorously contested. The answer was filed November 4, 1878, setting up alleged anticipations by Gray, Edison, Dolbear, and others. The record fills 1,200 printed pages; but before a decision was reached the case was compromised, and a decree entered by consent. The litigation ended at some time in the latter part of the year 1879. The last deposition was taken on the 19th of September in that year. The next contested suit was brought in the same court on the 28th of July, 1880, against Albert Spencer and others. An answer was filed in this case September 6, 1880, and depositions afterwards taken; some of those in the Dowd suit being used in this by stipulation. On the 27th of June, 1881, a decision was announced by Judge LOWELL sustaining the patent, upon which a decree was entered.[1] On the 14th of November, 1879, Abner G. Tisdel filed in the patent-office an application for a patent for "a new and useful improvement in speaking telephones," and on the 18th of November, 1879, Frank A. Klemm also filed an application for a patent for "a new and useful improvement in telephone transmitters." These inventions were transferred by assignment to Ernest Marx and Frank A. Klemm, of New York city, Moritz Loth, of Cincinnati, and Simon Wolf, of Washington. On the 6th of March, 1880, these parties entered into a mutual agreement to the effect that "each and all of their interests in said improvements and inventions, and the letters patent to be issued therefor, shall be merged and consolidated as common stock in a corporate body, under the laws of either of the states of Ohio, New York, or the general laws of the United States, relating to the formation of incorporations in the District of Columbia, or of such other states or territories as may be found necessary hereafter." This agreement was recorded in the patent-office March 10, 1880. On the 6th of May, 1880, Edgar W. Chellis, a merchant of Harrisburg, Pa., M. W. Jacobs, a lawyer at the same place, and Lysander Hill, a lawyer then residing in Washington, in the District of Columbia, made an arrangement with Daniel Drawbaugh by which they were to become jointly interested with him in his alleged telephone inventions, each to have a quarter interest. Nothing was paid for this, but each of the parties was to have one-fourth of anything that should be realized from the enterprise. On the 24th of May, 1880, Simon Wolf, one of the parties interested in the Klemm and Tisdel inventions, visited Harrisburg on business with Chellis in reference to telephone matters. On the 18th of May, four days before this visit, a patent was issued to Wolf and his associates upon the invention of Tisdel. While Wolf was in Harrisburg negotiations were begun with Chellis for a transfer of the Drawbaugh inventions to the owners of those of Klemm and Tisdel. These negotiations resulted in a conditional contract of the 22d of June, by reason of which Chellis, Jacobs, Hill, and Drawbaugh went to Washington; and there, on the 21st of July, 1880, Drawbaugh, claiming to "have invented certain new and useful improvements in the transmission of vocal speech, and the apparatus to be used for such purpose, for which I am about to make application for letters patent of the United States," assigned to Klemm, Marx, Wolf, and Loth "the full and exclusive right to the said invention, as fully set forth and described in the specification prepared and executed by me, dated the 21st day of July, 1880, preparatory to obtaining letters patent of the United States therefor;" and he at the same time, and by the same instrument, authorized and requested the commissioner of patents to issue the patent to his assignees, "each as assignee of one-fourth part." The specification referred to in the assignment has not been put in evidence in any of the cases. In the course of taking the testimony, it was called for by the Bell Company; but the counsel for the opposite party refused to produce either the original, or a copy from the patent-office. The assignment was recorded in the patent-office July 22, 1880, and in the official digest of assignments the following notation appears: "About to make appl'n. Spe'n dated July 21, 1880." On the morning of July 22, 1880, the following appeared in the Cincinnati Commercial, a newspaper printed at Cincinnati, Ohio:

"TELEPHONE COMBINATION. *Special to Cincinnati Commercial.* WASHINGTON, D. C., *July* 21. An application for a patent was filed to-day that, in consequence of its vastness of interest, as well as wealth of prospect, renders it a subject of national interest. A company of leading business men has been formed that has bought up all the telephone patents antedating those now in use, and known as the Bell, Gray, and Edison patents. The company is composed of leading business men from all parts of the country, Cincinnati being largely represented and interested. The cash capital of the company is $5,000,000, with head-quarters in New York, and in about sixty days they will open up the telephone, which will certainly result in the driving out of all telephones in the market, save the ones they hold, or else the compelling the Gray, Bell, and Edison lines to pay the new company a munificent royalty. It appears from the testimony now on file, and in the possession of the new company, which is conclusive and exhaustive, that the inventor of the telephone is a poor mechanic, living near Harrisburg, Pa., named Daniel Drawbaugh. Owing to his poverty, he was unable to push his patent on the market. The new company have secured, and are sole possessors of, this invention, antedating those now in use. They are also owners of four patents for telephones issued to Mr. Klemm, of New York. A large number of capitalists were here to-day to see the filing of the application, and they assert, with a positiveness that is almost convincing, that it will not be long till they have entire charge of the telephones, not only in this country, but in the world, and that they will be able to establish lines by which messages can be transmitted for almost a song. Mr. Lipman Levy, of the law firm of Moulton, Johnson & Levy, of Cincinnati, was here to-day, in the interest of the Cincinnati parties, who, as already stated, are among the most prominent financial men of our city."

Afterwards, on the 23d of August, 1880, the following appeared in the Journal of Commerce, a newspaper printed in the city of New York:

"A New Telephone Company. A company has recently been formed in this city with a capital of $5,000,000, for the purpose of manufacturing telephones. The company is to be known as 'The People's Telephone Company,' and a number of leading capitalists in this city and Cincinnati are interested in it. The telephones are to be manufactured under the patents of Frank A. Klemm and Abner G. Tisdel, and the application for patents of Daniel Drawbaugh, of Eberly's Mills, Cumberland county, Pa., filed July 21, 1880. It is claimed by those interested in the new enterprise that Drawbaugh is really the inventor of the telephone, and had completed one years before Professor Bell, or any one else, had manufactured one. He was, however, in very humble circumstances; and his neighbors who knew of his experiments looked upon him as a harmless lunatic. He continued improving his original telephone, and it is claimed that the one which the new company proposes to furnish is superior to any now in use. The company has fitted up a factory in Brooklyn, and in three months will be prepared to supply 1,000 of the new telephones. As soon as operations are actively commenced, it is expected that legal proceedings will be begun against the new company by the Gold & Stock Telegraph Company, which holds most of the existing patents, and a long and interesting legal fight is anticipated."

On the 30th of August, 1880, the People's Telephone Company was incorporated under the general laws of New York, with an authorized capital stock of $5,000,000, for "manufacturing, constructing, owning, furnishing, letting, and selling telephones, and the apparatus used therewith, under the inventions and patents of Abner G. Tisdel, Frank A. Klemm, Daniel Drawbaugh, and other inventions and patents which may hereafter be assigned to said company," and on the 4th of September, 1880, Klemm, Loth, Marx, and Wolf, in consideration of $4,999,550, represented by 99,991 shares of stock, assigned and transferred to that company all their interest in the Klemm, Tisdel, and Drawbaugh inventions; those of Drawbaugh being described as "the inventions in telephones made by Daniel Drawbaugh of Eberly's Mills, Cumberland county, in the state of Pennsylvania, for which application for patents was made on or about the 21st day of July, 1880, and which was assigned to us on the [twenty-] first day of July, 1880, as more particularly appears in a deed of assignment recorded in the United States patent-office in Liber W 25, page 85, in the Book of Transfers of Patents." For the assignment from Drawbaugh to Klemm, Marx, Loth, and Wolf, $20,000 was paid in money to Chellis, Jacobs, Hill, and Drawbaugh, and they were also to have a certain amount of the stock of the proposed corporation when formed. What amount they actually got, Chellis, who was sworn as a witness in the case, declined to tell, but he admitted it was large. At this time, and in this way, the attention of the general public was called for the first time to the fact that Drawbaugh claimed to have anticipated Bell in the discovery of the telephone. Bell's success had been proclaimed more than four years before at the Centennial Exposition in Philadelphia. In the mean time, inventions in aid of his discovery had been multiplied. According to the testimony of Park Benjamin, more than 100 patents had been issued and indexed under the word "Telephone." Numerous interferences had been declared and considered at the patent-office. Gray, Edison, Dolbear, and others had either claimed for themselves, or others had claimed for them, priority of invention and discovery, and Bell had thus far been sustained as against them all. Blake had perfected his microphone apparatus, and Bell's patent had become a great commercial success.

The People's Company either began, or threatened to begin, operations under its charter; and on the 20th of October, 1880, the Bell Company brought suit against it in the circuit court of the United States for the Southern district of New York to prevent any infringement of the Bell patents. In the bill it was alleged "that telephone exchanges now exist in more than two hundred and seventy-five towns and cities of the United States, and in every state thereof, and exist in substantially every city in the United States having more than 15,000 inhabitants, and in many smaller places;" "that there are now in use more than 100,000 electric speaking telephones licensed by and paying royalty to" the Bell Company; "that the owners of said Bell patents, and those who now are, or heretofore have been, licensed by them, have devoted great time and attention, and large sums of money, to the development of the telephone, and the introduction thereof into extensive use, and to the proper construction of the most suitable telephone lines and systems and telephonic appliances, and have constructed many thousand miles of telephone lines for use with telephones owned by" the Bell Company, "and licensed by it for such use; and that nothing which the defendants, or F. A. Klemm, A. G. Tisdel, and D. Drawbaugh, * * * have done, has contributed in any substantial way to the development of the telephone, or the introduction thereof into use." The bill then avers that Klemm, Marx, Loth, and Wolf, having become the owners of the Klemm and Tisdel improvements, and having heard that Drawbaugh "claimed that he had made some experiments relating to electric speaking telephones, (which experiments, if made, were incomplete, imperfect, unfruitful, and long before abandoned,) entered into an arrangement with him to set up and claim that he was the first inventor of the speaking telephone, and to make application for a patent therefor; and thereafter, alleging and pretending that said Drawbaugh was the original and first inventor of the electric speaking telephone, and that electric speaking telephones had not before such application been in public use or on sale for more than two years, with the knowledge and consent of Drawbaugh, they did, on or about the 21st day of July, 1880, induce him to make, and cause to be filed in the patent-office of the United States, an application for a patent to issue to them as assignees of the said Drawbaugh, as the first and original inventor of the electric speaking telephone, the said defendants well knowing at the time that electric speaking telephones had been in public use by" the Bell Company and its licensees "for more than two years before said application." It was then further alleged that, if Drawbaugh had ever made his pretended inventions, they "have not been by him, or any one claiming under him, introduced into public use, and that knowledge thereof has been withheld from your orators and the public, except so far as they have been disclosed within the three months last past by certain newsaper publications." To this bill the People's Company filed an answer in December, 1880, or January, 1881. The record does not show the precise date. In this answer it was said that Drawbaugh was "the original and first inventor and discoverer of the art of communicating articulate speech between distant places by voltaic and magneto electricity," and, that, "long prior to the alleged inventions by" Bell, Gray, and Edison, he, "then and now residing at Eberly's Mills, constructed and operated practical working electric speaking telephones at said Eberly's Mills, and exhibited their successful operation to a great number of other persons resident in his vicinity and elsewhere;" that his telephones, as then constructed and operated, "contained all the material and substantial parts and inventions patented" in the patents of Bell, and "also other important and valuable inventions in electric and magneto telephony, and were fully capable of transmitting, and were actually used for transmitting, articulate vocal sounds and speech between distant points by means of electric currents; that some of the original machines and instruments invented, made, used, and exhibited to many others long prior to the said alleged inventions of Bell, or either of them, are still in existence, and capable of successful practical use, and are identified by a large number of persons who personally tested and used them, and knew of their practical operation and use, in the years 1870, 1871, 1872, 1873, 1874, and both prior and subsequent thereto; that certainly more than fifty, and probably not less than one hundred, persons, or even more, were cognizant of said Drawbaugh's invention and use of said telephones, and of his claim to be the original and first inventor thereof prior to the alleged inventions of said Bell, or either of them; that said Drawbaugh, for more than ten years prior to the year 1880, was miserably poor, in debt, with a large and helpless family dependent on his daily labor, and was from such cause alone utterly unable to patent his invention, or caveat it, or manufacture and introduce it on the market; that said Drawbaugh never abandoned his said invention, nor acknowledged the claims of any other person or persons thereto, but always persisted in his claims to it, and intended to patent it as soon as he could procure the necessary means therefor; that said Drawbaugh never acquiesced in the public use of said Bell, Gray, Edison, Blake, or other telephones; nor in the claims of the alleged inventors thereof, nor gave his consent to such use." It is then said that Drawbaugh, after finding by experiment that his invention was capable of successful working, "conceived that its range and capacity for usefulness to the public might be very greatly enlarged; that many improvements of great value might be made and added to it, which, without departing from its principle, might increase its value to himself and to the public, and therefore set himself at work to discover and invent such improvements; that he discovered and invented some of said additional improvements prior to any alleged invention by Bell; and that notwithstanding his embarrassed and impoverished pecuniary condition, and his utter want of proper mechanical tools, materials, and appliances to conduct such work, he labored with all reasonable diligence to perfect and adapt his said improvements, and did finally, in due exercise of such reasonable diligence, perfect and adapt the same; and that in so far as the said Bell has incorporated such improvements in his said two patents, or either of them, he (the said Bell) has surreptitiously and unjustly obtained a patent or patents for that which was in fact first invented by Drawbaugh, who was using reasonable dilligence in perfecting and adapting the same, and therefore the patent or patents of the said Bell therefore is or are invalid and void." It is then said that "the defendant in good faith, and relying upon its legal rights, * * * caused applications to be made and filed in the patent-office for letters patent on the inventions of the said Daniel Drawbaugh, with the intention of procuring interference proceedings to be instituted, in accordance with the statute, against the patents of said Bell, and the pending applications of said Gray, Edison, and others, in order that said Drawbaugh may be adjudged by the commissioner of patents to be, as he rightfully is, the original and first inventor of the electric speaking telephone, and may be adjudged entitled to receive a patent or patents therefor." The People's Company began taking depositions on the 19th of April, 1881, but Drawbaugh himself did not appear as a witness until December 7, 1881. After that time others were examined, and when the proofs were closed between three and four hundred witnesses had been produced whose testimony was taken, and put into the record, to establish the priority of Drawbaugh's invention. This testimony, as is now claimed, shows the story of that invention to have been as follows: "Early conception and experiments with the continous current, 1862, 1866, and 1867. Teacup transmitter and receiver, 1866 and 1867. Tumbler and tin-cup and mustard can, ('F' and 'B,') 1867 and 1869. Improvement on 'B' ('C,') 1869, 1870. Further improvement upon 'C,' and the more perfect magneto instrument, 'I,' 1870, 1871. Mouth-piece changed to center and adjusting screw inserted, (Exhibit A,) 1874. 'D' and 'E,' perfectly adjusted and finished magneto instruments, January and February, 1875. 'L,' 'M,' 'G,' and 'O,' from February, 1875, to August, 1876. 'H,' August, 1876. 'J,' 'N,' and 'P,' 1878." This statement of the Drawbaugh claim we have quoted from the brief of counsel appearing in his behalf, and his success in the litigation has been placed, as we understand it, both in the answer and in the argument, on the truth or falsehood of what is thus set forth. The letters "F," "B," etc., in the statement, refer to exhibits in the cause, being certain instruments claimed to have been made and used by Drawbaugh in the progress of his work, and preserved until now. The original tea-cup instrument was not produced, but Drawbaugh in his deposition gave what he said was a drawing, showing how it had been constructed. "F," "B," "C," "I," and "A" were neither of them in a condition for use when they were put in evidence, and no one of all the witnesses except Drawbaugh could tell how they were originally constructed, or what the process was by which sound was transmitted when they were used. All any of the witnesses could say on that subject was that they had used one or more of the different instruments at Drawbaugh's shop; had heard sounds, and sometimes spoken words, through them; and that Drawbaugh told them the sound was carried on the wire by electricity. There was nothing whatever produced in print or in writing on the subject; not even a memorandum or a drawing of any kind. And there is nothing in the testimony to show that Drawbaugh ever told any one how his earlier instruments were made, or what his process was, until he was called as a witness in December, 1881, and explained it in his testimony. This was nearly twenty years, according to the present claim, after he had begun his experiments; nearly seven after he had made and used "D" and "E," perfectly adjusted and finished magneto instruments; and more than five after "L," "M," "G," "O," and "H" had been constructed, and kept in his shop. It was also nearly six years after the date of Bell's patent, more than five after the success of his discovery had been proclaimed at the Centennial Exposition in Philadelphia, four after his process had got into public use, three after it had become an established success, and two after he had brought his first suit for the establishment of his rights against Dowd, who represented the Western Union Telegraph Company, to a successful termination.

Under these circumstances, it becomes important to consider the conduct of Drawbaugh in reference to his alleged invention during this 20 years of eventful history, as connected with the discovery and use of telephones. If his present claim is true, his experiments began almost as far back as those of Reis, and he had in his shop at Eberly's Mills, within three miles of Harrisburg, telephones that were substantially perfect months before Bell, on the 2d of June, 1875, got the clue to his subsequent discoveries. It is conceded that "D" and "E," made, as is claimed, in February, 1875, are substantially as good magneto instruments as any Bell had used before December, 1881; and "L," "M," "G," "O," and "H," all of which it is claimed were constructed by August, 1876, and some in February, 1875, are as good, or nearly as good, microphones as those of Blake, which were not invented until 1878. This is the theory of Drawbaugh's defense as it is set forth in the answer and in the argument, and by it his case must stand or fall. The claim is that the discovery of the process was complete, and that perfect telephones had been made, and were in a condition for use, a year and more before Bell got his patent. Drawbaugh was, when he gave his deposition, 54 years of age, and had lived all his life at or near Eberly Mills, a small village near Harrisburg. He was a skillful and ingenious mechanic; and if he made "D" and "E," and the instruments which came after them, at the time it is said he did, he had good tools and good materials in 1875 and 1876, and was capable of doing the best of work. He was also somewhat of an inventor, and had some knowledge of electricity. According to the testimony, he was an enthusiast on the subject of his "talking machine," and showed it freely to his neighbors and people from the country when they visited his shop. The Centennial Exposition was opened at Philadelphia in May, 1876, and Drawbaugh visited it on the 17th of October, 1876, remaining four or five days. Before he went he had heard, as he says, that some one besides himself had invented a speaking telephone, which he had the impression was on exhibition there. If what he now claims is true, he had then on hand in his shop Exhibits D, E, L, M, G, O, and H; all of them good instruments of their kind, and capable of transmitting speech, and some of them but just finished. Bell's apparatus had been exhibited to the board of judges in June before, and had attracted marked attention. The matter was much discussed in the public press, and yet it never seems to have occurred to Drawbaugh to take any of his telephones with him when he went, although they were small in size, and some, or all of them, could have been carried without serious inconvenience.

When giving his testimony, he was examined in chief as to that visit; and this is what he said on the subject of telephones: "*Question* 386. Did you attend the Centennial Exhibition at Philadelphia in the year 1876? *Answer.* Yes, sir; I did. *Q.* 387. Can you give the date on which you went there? *A.* I can by reference to a book. It was October 17, 1876. The 17th was a day on which I dated a letter from Philadelphia, while I was there on that visit. *Q.* 388. How long did your visit there last? *A.* About four or five days, to the best of my recollection. *Q.* 389. Who went with you on that visit? *A.* Mr. George Leonard. *Q.* 390. Was that the only visit to the Centennial Exhibition that you made? *A.* Yes, sir; it was. *Q.* 391. At the time that you went there, or before that time, had you heard that somebody else besides yourself had invented a speaking telephone, or a telephone? *A.* Yes, sir; some time before that. I don't remember how long, but not a great while. *Q.* 392. When you went there, did you suppose it would be on exhibition there? *A.* I don't remember whether I had heard that it was on exhibition or not; but I got the impression some way that it was on exhibition. *Q.* 393. While you were there at the Centennial, did you see any telephones, or make an effort to see any there? *A.* Yes, sir; I made an effort, and seen an instrument called a 'telephone,' and supposed it to be the instrument spoken of,—the one of which I had heard. I was looking and had made some inquiry, and was directed or came to a portion of the building where I saw on a counter some man's telephone; the name I don't remember. At that time, or several times that I called, there was no one there to attend to it. I spoke to another party that had something else on exhibition—I don't recollect what it was—just near by, and I asked him whether there was any one there to attend, or to show the instruments. I was informed, then, there was no one there to show them. *Q*. 394. If you remember, please state what kind of an instrument it was that you saw there, and state what information you were able to obtain there regarding it, and its mode of operation. *A*. There was a number of instruments placed onto a raised portion,—something like a shelf; that is, it resembled something like pigeon-holes, a box open in front, and each instrument at the back of it had an electro-magnet. The number of instruments I don't remember. I don't remember of counting them. If I am not mistaken, there may have been a dozen or more, perhaps. Some were larger than others. I could not give you a much better description than that. I couldn't get any information about them. This attendant made some remarks about the instruments; but he didn't understand them, and couldn't explain them. I was several feet from where the instruments were. They were placed—it occurs to me—on a raised place like a shelf, just about high enough for a man to speak into; that is the way it looked to me. I did not go in behind the counter to examine them, although there was an opening to go in by, because I did not like to make too free, as there was no one there. *Q*. 395. Did you see any circulars lying around there referring to these instruments, or other advertisements of them? *A*. I don't remember about that; it may have been. *Q*. 396. What was your impression as to the character of the instruments when you finally left them? *A*. I was impressed with the idea that they were instruments to telegraph by sounds. A certain sound to represent a certain letter of the alphabet. I am not certain how I got the idea, or whether any person told me that at the time, but that is the idea that I had. When I said certain sounds, I meant that sounds of a different pitch would represent different letters. *Q*. 397. Do you know whether that was 'Gray's Harmonic Telegraph' that you saw there or not? *A*. It didn't say 'Telegraph;' I am confident it was called 'Telephone.' I didn't see the working parts of the interior, except the electro-magnets. I took the name of the man and his address on a piece of paper, and put it in my pocket, but I don't know what became of it. I don't know whether it was 'Gray's Harmonic Telegraph' or not. *Q*. 398. Did you see any tuning forks about it? *A*. I did not."

That was all he did during his entire visit to ascertain whether any one besides himself had actually entered upon this then new and interesting field of invention and discovery. He spoke to no one about what he had done himself, and he made no special effort to find out whether that which was on exhibition was in any respect like what he had at home. Neither did he when he got home, so far as the records show, say anything to his neighbors or visiting friends about what he had seen or heard. He had apparently lost all interest in "talking machines." Not so, however, with his other inventions. The testimony shows that during the early part of 1876 he was much occupied in building an electric clock, which he thought of exhibiting at the Centennial. This he did not do, however, but either just before he went to Philadelphia, or soon after, Rufus E. Shapley, a jeweler of Mechanicsburg, went by his invitation, or on his suggestion, to Eberly's Mills to look at the clock which he had made. Soon afterwards the clock was taken to Shapley's store in Mechanicsburg, and on the 8th of November, 1876, Drawbaugh, by an instrument in writing, transferred to Shapley a half-interest in the "clock I am getting up, the said R. E. Shapley to pay for patenting the same." Shapley had then $2,000 in money which Drawbaugh was anxious to have him invest in that business, and the clock was taken by him to his shop so that it might be examined with that end in view, if it should prove to be useful. Some time afterwards it was taken back to Eberly's Mills, where it remained until April 1, 1878, or thereabouts, when a clock company was formed, and that clock, or another one substantially like it, was taken about the country for exhibition. For this Drawbaugh was paid $500, with an interest in the profits, and on the 20th of September, 1878, he applied for a patent for "improvement in earth batteries for electric clocks," which was issued January 14, 1879, to the members of the clock company. The enterprise does not seem to have been productive of any great success. In November or December, 1878, while this clock was on exhibition at Harrisburg, Drawbaugh was introduced to Edgar W. Chellis. He had with him at the time a "wooden model of a faucet" that he wanted Chellis and another man to take each a third interest in. An arrangement was afterwards made by which Chellis got a two-thirds interest, he paying for it $250 January 7, 1879. On the 14th of the same month Drawbaugh filed in the patent-office an application for a patent for "improvement in rotary measuring faucets;" Chellis to have a two-thirds interest. After this application an interference was declared, March 29, 1879, between Drawbaugh and David A. Hauck, who had filed a conflicting application January 17th. In his preliminary statement upon this interference, Drawbaugh said that he had conceived the idea of his faucets, and sketched them, late in the fall of 1876; that he made a working model in the spring of 1877, and actually tested it then, but the patent-office model was not completed until about the 1st of November, 1878. The case was closely contested, but finally decided in favor of Drawbaugh, January 15, 1880. The patent was granted to him and Chellis, July 6th of the same year. In this contest Jacobs and Hill, who afterwards became interested in his telephone claims, appeared as the counsel of Drawbaugh. On the 2d of July, 1879, Drawbaugh filed another application in the patent-office for "improvement in water motors;" Chellis to have in this, also, a two-thirds interest. Upon this application a patent was issued March 16, 1880. It is impossible to believe, if Drawbaugh had in his shop, when he reached home from the Centennial, Exhibits D, E, L, M, G, O, and H, or even D and E alone, that he would have set himself to work, in the first instance, at developing his clock enterprise, or perfecting his former conception of a measuring faucet, instead of making some effort to call the attention of his friends to his great discovery of the telephone, which he was in danger of losing by the patent which had been issued to another, and which he could not but have known was even then attracting the greatest attention. And in this connection it must be kept in mind that the theory of the defense is, as stated in the answer, that Drawbaugh had at that time fully perfected his invention; and that while, at first, he "conceived that its range and capacity for usefulness to the public might be very greatly enlarged," he had, before the date of Bell's patent, "notwithstanding his embarrassed and impoverished pecuniary condition, and his utter want of proper mechanical tools," finally perfected his work. His conduct afterwards, therefore, is to be judged, not as that of one who was still in the midst of his experiments, and doubtful of the results, but of one who had arrived at the end and had completed his success.

No man of his intelligence, with or without the enthusiasm upon the subject which it is said he possessed, could have remained silent under such circumstances. As we have read the testimony, it is not even pretended that he took any of his instruments outside of his own village until May, 1878, when, as is claimed, he showed one to his friend Stees, in Harrisburg, whom he had known for years, and who was the first to use, and, in fact, was then using, a Bell telephone, in that place, upon a private line of his own between his office and his shops. This produced no results; and when afterwards, in January, 1879, Chellis was told that Drawbaugh had "a phonograph and a telephone that he had invented," he gave it no attention, because, to use his own language, "I was interested in the faucet and motor business, and wished to push them, and I did not think we could do much with the telephone, as Bell had a patent, and I did not know that he could antedate them." And again, when speaking of a conversation he had with Drawbaugh, he said: "I advised him to drop it,—the telephone,—as he could not antedate Bell. He said he did not know about that; that he had been working on it a good while. It was his way of expressing himself. When I would say, 'You can't antedate Bell,' he would say, 'I don't know about that; I have been working at it a good while.'" This, it must be remembered, was in 1879, after the telephone had become a success, and after it had been a year or more in use in Harrisburg, where Chellis lived. It is impossible to believe that either Chellis or Drawbaugh was ignorant of the approximate time of Bell's invention, which had been the subject of frequent newspaper comment from the time of its exhibition at the Centennial. The subject was often referred to in the Harrisburg and Mechanicsburg papers, and it is not for a moment to be supposed that all of these various articles escaped their attention. Under such circumstances, if it were true that Drawbaugh had made his "D" and "E," as is now claimed, in February, 1875, he certainly would have said so, and would not have contented himself with so doubting an answer to Chellis' suggestion of his inability to antedate Bell as that which Chellis now says he gave.

Another important fact in this connection is one which is proved by the testimony of Andrew R. Kiefer, who from 1863 had been division telegraph operator, having charge of the middle division, of the Pennsylvania Railroad, and residing in Harrisburg. From 1867 to the winter of 1881–82 he was a member of a partnership firm in that place which was engaged in "the manufacture of burglar alarms, electric hotel annunciators, and fine electric work for the government,—instruments for the signal bureau, patent models, etc." He had also, since 1876, kept a place for the sale of electrical supplies. He had known Drawbaugh certainly since 1876, and probably before. Drawbaugh met him on different occasions and talked upon electrical matters. In the course of their acquaintance, Drawbaugh showed him an electrical fire-alarm apparatus, and the works of his electric clock; but the subject of telephones was never alluded to between them until in the summer of 1881, when this occurred. We quote from Kiefer's deposition: "In the summer of 1881, I took my wife out for a drive, and went over to see his (Drawbaugh's) works, never having seen them, and having promised to come and see him some time. My wife, not caring about going through the shop, remained in the carriage, and I went through alone with Mr. Drawbaugh. He showed me through the shops, and introduced me to Mr. Chellis, and showed me parts of the water motor and some other things of his getting up. On account of my wife's being in the carriage alone, I did not stay long. As I stepped into, or was just in, the carriage, Mr. Drawbaugh said, 'I forgot to show you my telephone.' I did not get out again to go and see it, and I drove away without seeing it, expecting to see it again; but I have never got over to the shop since." This was after the suit of the Bell Company against the People's Company was begun, and of course after the matter got into the hands of Chellis and his associates. It is no answer to the criticism of Drawbaugh's conduct in this particular to say, as was said in argument, that "one reason why he did not speak or apply to every man with whom he had personal acquaintance was that he was ridiculed by his neighbors; that his invention was considered a humbug by them, and of no commercial value." Bell's success was proclaimed in the Harrisburg Patriot as early as February 26, 1877, and the days of ridicule were then past. If Drawbaugh had at that time in his shop the machines which it is now claimed were all complete as they now are by August, 1876, and most of them before, there cannot be a doubt that he would have taken them to some place where they could be tried, and show that they would do what he had all along claimed for them. All he had to do, at any time after he came back from the Centennial, was to take any pair of his little instruments to his friend Zeigler or his friend Stees at Harrisburg, attach them to a line wire, and show what he had. They were men who could appreciate his achievement, and help him if it was, as he now says it was, a success. It would certainly have been easier then, within two years of the time the first of them were made, and within a year of the date of Bell's patent, to show that he "antedated" Bell, than it was three years afterwards, when he was brought into the controversy through the instrumentality of his associates; not, as must be evident to all, to get a patent for himself, but to defeat that of Bell. And in this connection it is specially significant that the application which it is claimed was made for a patent on the 21st of July, 1880, and the specification of his invention which was then written out, have been purposely and designedly kept out of the case, although their production was demanded. They were written before this suit was begun, and it is impossible to believe that they would have been withheld, at least upon the call of the opposite party, if they were in all respects consistent with the subsequent developments of the case. The excuse given by counsel at the time, that they were "in the secret archives of the patent-office," and, "if produced and published in this cause, would possibly invite the filing of contesting applications, and result in interference and additional litigation, besides unnecessarily prolonging the taking of testimony here, and increasing the expenses," we cannot accept as satisfactory, especially as in the answer it was said that one object of filing the application was to procure "interference proceedings to be instituted against the patents of Bell, in order that Drawbaugh may be adjudged by the commissioner to be, as he rightfully is, the original and first inventor."

We have not overlooked the depositions that have been taken in such large numbers to show that Drawbaugh was successful with "F," "B," "C," "I," and "A," before "D" and "E" were made. They have been studied with care, and, if they contained all the testimony in the case, it would be more difficult to reach the conclusion that Drawbaugh's claim was not sustained. But in our opinion their effect has been completely overcome by the conduct of Drawbaugh, about which there is no dispute, from the time of his visit to the Centennial until he was put forward by the promoters of the People's Company, nearly four years afterwards, to contest the claims of Bell. He was silent, so far as the general public were concerned, when, if he had really done what these witnesses now think he did, he would most certainly have spoken. There is hardly a single act of his connected with his present claim, from the time he heard, before going to Philadelphia, that some one else had invented a telephone which was on exhibition at the Centennial, that is not entirely inconsistent with the idea, even then, of a complete discovery or invention by himself which could be put to any practical use. It is not pretended that what he did was done in private. He had influential friends, with ample pecuniary resources, ready to help him in bringing out his inventions when they promised success. He easily got aid for his clock and for his faucet. The news of Bell's invention spread rapidly and at once, and it took but a few months to demonstrate to the world that he had achieved a brilliant success. If it were known at Eberly's Mills alone that Drawbaugh had been doing the same thing for years in his shop there,—and it certainly would have been known all through the little village if it had actually been done,—no one can believe that the public would be kept in ignorance of it until four years afterwards, when a "special" from Washington "to the Cincinnati Commercial" announced a "telephone combination" "to have entire charge of the telephones, not only in this country, but in the world," that could transmit messages "for almost a song."

But there is another fact in this case equally striking. As has already been seen, "F," "B," "C," and "I" were in no condition for use when they were produced and put in evidence. They were mere "remains," and no one but Drawbaugh himself could tell how they were made, or how they were to be used. He undertook to reproduce some of them, especially "F" and "B." This was in the latter part of 1881, while the testimony was being taken. The Bell Company proposed that they should be tried to see if they would do what he witnesses said had been done with the originals, which the "remains" how must have been exceedingly primitive in their character. The testimony also shows that, when they were originally used by or in the presence of the witnesses, no particular care was taken in their adjustment. They were lying around in the shop, or standing upon shelves. Some say that when experiments were made they were held in the hand, or allowed to stand on the table. Many testify to satisfactory results, and Drawbaugh himself said in his deposition: "I would have persons in the cellar reading printed matter,—some advertisement or something,—and I could hear the words that were read; and at other times I would go down into the cellar and read something, and coming up they would repeat the words to me that I had read."

The proposition of the Bell Company was accepted, and the reproductions were tried in March, 1882, under the most favorable circumstances. Three days were occupied in the test, and it is substantially conceded that it was a failure. Occasionally a sound was heard, and sometimes a word, but "it would not transmit sentences." At the time of these experiments, "F," which was the transmitter, was placed on a table, and used as Drawbaugh said it was originally. Two years afterwards other reproductions were presented, differently constructed, and used in a different way, and these would "talk;" but they were neither made nor used in the same way as the originals. To our minds, the result of the second experiments conclusively showed that the original instruments could not have done what the witnesses supposed they did, and that what they saw and heard was produced by some other means than an electric speaking telephone. We do not doubt that Drawbaugh may have conceived the idea that speech could be transmitted to a distance by means of electricity, and that he was experimenting upon that subject; but to hold that he had discovered the art of doing it before Bell did, would be to construe testimony without regard to "the ordinary laws that govern human conduct." *Atlantic Works* v. *Brady*, 107 U. S. 192, 203, 2 Sup. Ct. Rep. 225. Without pursuing the subject further, we decide that the Drawbaugh defense has not been made out.

Another objection to Bell's patent, put forth in the oral argument of Mr. Hill, and in the printed brief signed by him, and in that signed by Mr. Dixon, is that his application as originally filed in the patent-office did not contain his present fourth claim, or any description of the variable resistance method, and that all which now appears in the specification on that subject, including the fourth claim, was surreptitiously interpolated afterwards. Bell's application was filed February 14, 1876, and afterwards, during the same day, Elisha Gray filed a *caveat*, in which he claimed as his invention "the art of transmitting vocal sounds or conversations telegraphically through an electric circuit," and in his specification described the variable resistance method. The precise charge now made in the printed brief of Mr. Hill is that "Mr. Bell's attorneys had an underground railroad in operation between their office and Examiner Wilbur's room in the patent-office, by which they were enabled to have unlawful and guilty knowledge of Gray's papers as soon as they were filed in the patent-office," and "that an important invention, and a claim therefor, were bodily interpolated into Bell's specification, between February 14, 1876, and February 19, 1876, by Pollok, in consequence of the guilty knowledge which the latter already had of the contents of Gray's *caveat* before the declaration of interference with Gray on February 19th." So grave a charge, made in so formal a manner, is entitled to careful consideration. It involves the professional integrity and moral character of eminent attorneys, and requires us to find from the evidence that after Bell swore to his application on the 20th of January, 1876, and after the application thus sworn to had been formally filed in the patent-office, an examiner, who got knowledge of the Gray *caveat* put in afterwards, disclosed its contents to Bell's attorneys; that they were then allowed to withdraw the application, change it so as to include

Gray's variable resistance method over Bell's signature, and over the jurat, and then restore it to the files, thus materially altered, as if it were the original; and all this between February 14th and February 19th. Although much stress was laid in argument on the fact that what purported to be a certified copy of the specification of Bell, as found in the file-wrapper and contents printed in the Dowd case, differed materially from the patent, the cause of these differences has been explained in the most satisfactory manner; and we entertain no doubt whatever that the specification as now found in the patent is precisely the same as that on which the order to issue was made. If any alterations were made, it was all done before February 19th; and the fair copy which is now found on the files of the office is precisely as it was when the order for the patent was granted. Not a shadow of suspicion can rest on any one growing out of the misprint of the specification in the Dowd case. All that remains, therefore, on which to rest this serious charge is that in a paper handed by Bell to George Brown, of Toronto, describing his invention, and which was intended to be used in England to secure a British patent, what is now claimed to be an interpolation in the American application is not to be found. It is but right to say that, during the whole course of the protracted litigation upon the Bell patent, no argument was ever presented based on this discrepancy until the brief of Mr. Hill was filed in this court on the 18th of January, 1887, six days before the argument in these appeals was begun. So far as we are advised, nothing had ever before occurred in the cases that seemed to make it necessary to prove when the variable resistance method or the fourth claim was put into the American application, or why it was left out of the paper handed to Brown. It seems always to have been assumed, until the cases got here, that, because it was in the American patent, it was rightfully there. Certainly, there is nothing in the pleadings in any of the cases to direct attention to the materiality of this fact. A comparison of the paper handed Brown with the American application shows that they differ in more than 30 different places, besides those which relate to the variable resistance method and the fourth claim. The differences are generally in forms of expression; thus indicating that one was written after the other, and evidently for the purpose of securing greater accuracy. The paper handed Brown was clearly a rough draft, and not a fair copy, for the record shows that it bore on its face the evidence of many erasures and interlineations. Bell says in his testimony that he began writing his specification in September or October, 1875, and wrote and rewrote it a number of times, finally adopting that mode of expression which seemed to him the best to explain his invention, and the relation which one portion bore to another. He visited Brown in Canada in September, and again in December, 1875. The arrangement was made between them on the 29th of December, at this last interview, by which Brown was to interest himself in getting out British patents. Other inventions besides the telephone were included in the contract entered into for that purpose. Bell returned to Boston on the 1st of January, and immediately set himself to work to complete his specification. He had it done so that it was taken to Washington by Mr. Hubbard about the 10th of that month, and delivered to Pollok & Bailey, the attorneys. It was then examined by the attorneys, found correct, and a fair copy made, and returned on the 18th to Bell in Boston for his signature and oath. It was signed and sworn to in Suffolk county, Mass., January 20th, and immediately returned to the attorneys. Afterwards Pollok met Bell in New York, and it was again gone over with care by the two together. No change whatever was made in it at that time, and Pollok took it back with him to Washington. On the 25th of January, 1876, Bell met Brown, who was then on the way to England, in New York. It is now assumed that the paper which Brown took to England was handed to him then; and, because the variable resistance method and the fourth claim were not in that, it is argued that they could not have been in the American specification at that time. But no one has said when the paper was actually handed to Brown. Bell says he cannot tell, but that it must have been after he made his contract with Brown on the 29th of December. As the American specification was signed and sworn to five days before the interview with Brown on the 25th of January, and the paper of Brown differs from it in so many particulars besides that now in question, it would seem to be clear that the paper was a copy of some former draft which Bell had made,—possibly, one taken to Canada in December,—and not of that which was perfected afterwards. As the specification which had been prepared and sworn to was a fair copy, without erasures or interlineations, the fact that the paper handed Brown was not a fair copy would imply that it was not intended to be an exact transcript of the other. At any rate, the bare fact that the difference exists, under such circumstances, is not sufficient to brand Bell and his attorneys, and the officers of the patent-office, with that infamy which the charges made against them imply. We therefore have no hesitation in rejecting the argument. The variable resistance method is introduced only as showing another mode of creating electrical undulations. That Bell had had his mind upon the effect of such a method is conclusively established by a letter which he addressed to Mr. Hubbard on the 4th of May, 1875, and which is found in the Dowd record, introduced into the Overland case by stipulation. Its insertion in his final draft of his specification is another proof of the care with which his work had been done.

In the case of the Clay Commercial Company objection was made to the sufficiency of the proof of the incorporation of the American Bell Telephone Company, and of its title to the Bell patents. Upon the first point the proof was (1) a special act of the general court of Massachusetts, entitled "An act to incorporate the American Bell Telephone Company,"[1] which authorized certain persons therein named, and their associates, to organize themselves under the provisions of chapter 224 of the Acts of 1870,[2] and the acts in amendment thereof, for telephone purposes; and (2) a certificate of the secretary of the commonwealth, in the form required by section 11, of chapter 224, that certain persons, among whom were the most of those mentioned in the special act, were legally organized and established as an existing corporation, under the name of the American Bell Telephone Company. This section made such a certificate "conclusive evidence of the existence of a corporation" organized under that chapter. The authority granted by the special act to the persons named to organize as a corporation in this way gave them the authority to select a corporate name, and also made the statutory certificate conclusive evidence of their corporate existence. The objections to the proof of title are not, in our opinion, well taken. We do not deem it necessary to add to the length of this opinion by referring particularly to the testimony on that point.

This disposes of all the cases so far as the patent of March 7, 1876, is concerned. It remains only to consider the patent of January 30, 1877, about which but little has been said either in the oral or printed arguments. Apparently, it received but little attention by counsel or the court in either of the cases below. In the Dolbear case, it was, by consent, excluded from the decree, and of course is not presented by that record in this court. In all the other cases the patent was sustained, and the Clay Commercial Company was adjudged to have infringed the third, fifth, sixth, seventh, and eighth claims; the Molecular Company, the sixth, seventh, and eighth, but not the fifth; the People's Company, the fifth, sixth, and eighth; and the Overland Company, the third, fifth, sixth, seventh, and eighth. From the decree in favor of the Molecular Company as to the fifth claim the Bell Company has appealed.

[1]Approved March 19, 1880, (Laws & Resolves Mass. 1880–81, c. 117, p. 74.)
[2]Laws & Resolves Mass. 1870–71, p. 154.

In the case of the Clay Commercial Company, it was alleged in the answer that the substantial and material parts of the things described and claimed were described and claimed in a prior British patent taken out by or for Bell, dated December 9, 1876, and that, inasmuch as the American patent does not bear the same date with the foreign patent, and is not limited to expire therewith, it is void. This patent has not been pressed in the argument here, and in our opinion it has been settled by the decision of this court in *O'Reilly* v. *Morse*, 15 How. 62, 112, and impliedly by that in *Siemens* v. *Sellers*, 123 U. S. 276, 8 Sup. Ct. Rep. 117, (at the present term,) that the effect of section 4887, Rev. St. is not to render invalid an American patent which does not bear the same date as a foreign patent for the same invention, but only to limit its term. The patent itself is for the mechanical structure of an electric telephone, to be used to produce the electrical action on which the first patent rests. The third claim is for the use in such instruments of a diaphragm, made of a plate of iron or steel, or other material capable of inductive action; the fifth, of a permanent magnet constructed as described, with a coil upon the end or ends nearest the plate; the sixth, of a sounding box, as described; the seventh, of a speaking or hearing tube, as described, for conveying the sounds; and the eighth, of a permanent magnet and plate combined. The claim is not for these several things in and of themselves, but for an electric telephone, in the construction of which these things, or any of them, are used; hence the fifth claim is not anticipated by the Schellen magnet, as was decided in the Molecular case below. The patent is not for the magnet, but for the telephone of which it forms but part. To that extent the decree in that case was erroneous.

It follows that the decree in each of the cases, so far as it is in favor of the Bell Company, and those claiming under it, must be affirmed; and that the decree in the Molecular case, so far as it is against that company on the fifth claim of the patent of January 30, 1877,[1] must be reversed, and a decree directed to that extent in its favor. It is consequently so ordered.

GRAY, J., was not present at the argument, and took no part in the decision of these cases. LAMAR, J., not being a member of the court when these cases were argued, took no part in their decision.

BRADLEY, J., (*dissenting.*) Mr. Justice FIELD, Mr. Justice HARLAN, and myself are not able to concur with the other members of the court, sitting in these cases, in the result which has been reached by them. Without expressing an opinion on other issues, the point on which we dissent relates to the defense made on the alleged invention of Daniel Drawbaugh, and applies to all the cases in which that invention is set up. We think that Drawbaugh anticipated the invention of Mr. Bell, who, at most, is not claimed to have invented the speaking telephone prior to June 10, 1875. We think that the evidence on this point is so overwhelming, with regard both to the number and character of the witnesses, that it cannot be overcome. As this is a question of fact, depending upon the weight of the evidence, and involves no question of law, it does not require an extended discussion on the part of those who dissent from the opinion of the majority; which is very ably drawn, and presents the case with great clearness and force. On the point mentioned, however, we cannot concur in the views expressed.

The essence of the invention claimed by Mr. Bell is the transmission of articulate speech to a distance, by means of an electrical current subjected to undulations produced by the air vibrations of the voice. There are two modes (as yet discovered) by which these undulations may be thus produced. In one, they are produced by interposing in the circuit a substance whose elec-

[1] 32 Fed. Rep. 214, 23 Blatchf. 262.

trical conductivity may be varied by the concussions or vibrations of the air produced by the voice. This is called the "variable resistance process," because the electrical current is subjected to the variable resistance (or conductivity) of the substance thus interposed. By the other mode, the undulations are produced by the inductive effect of an armature, (or small, flat piece of iron,) attached to the membrane spoken against, and placed near to the poles of an electro-magnet situated in the circuit. In both cases the undulations impart the vibrations which caused them to another diaphragm at a distance (called the receiver) by means of an electro-magnet in the circuit, placed near to an armature affixed to such diaphragm. These vibrations, thus reproduced, are detected by the ear, and the spoken words are heard. We are satisfied, from a very great preponderance of evidence, that Drawbaugh produced, and exhibited in his shop, as early as 1869, an electrical instrument by which he transmitted speech, so as to be distinctly heard and understood, by means of a wire, and the employment of variable resistance to the electrical current. This variable resistance was produced by causing the electrical current to pass through pulverized charcoal, carbon, and other substances, acted upon by the vibrations of the voice in speaking. This was the whole invention, so far as the principle of variable resistance is concerned. We are also satisfied that as early as 1871 he reproduced articulate speech, at a distance, by means of a current of electricity, subjected by electrical induction to undulations corresponding to the vibrations of the voice in speaking,—a process substantially the same as that which is claimed in Mr. Bell's patent.

In regard to the instrument in which the principle of variable resistance was used, more than 70 witnesses were examined, who either testified to having seen it and heard it, or established such facts and circumstances in relation to it as to put its existence and date beyond a question. With regard to the instrument in which electrical induction was employed to produce the requisite undulations, some 40 or 50 witnesses were produced, many of whom saw it, and heard speech through it; and others either saw it, or heard it talked about in such a manner as to fix the time when it was in existence. On the questions of time and result, there is such a cloud of witnesses in both cases, that it seems almost impossible not to give credence to them. The evidence of some of them may have been shaken with regard to the time they had in mind; but that of the great majority was not shaken at all, but corroborated by circumstances which rendered the proof irrefragable. Many of them, it is true, were plain country people; but they heard the words through the instrument, and that is a matter about which they could not be mistaken. It did not require science nor learning to understand that. But the witnesses were not confined to this class. A number of them were people of position in society, official, professional, and literary; all, however, like the inventor, regarding the matter more as one of curiosity than of public importance. As it would serve no useful purpose to repeat the testimony of these witnesses, we shall refrain from doing so. We will only add that nearly all the original instruments used by Drawbaugh were produced on the trial, and identified by the witnesses. Some of them were broken, and in a dilapidated condition, but sufficiently perfect to be accurately reproduced. Their very form and principle of construction showed that they were intended for speaking telephones, and nothing else. Drawbaugh certainly had the principle, and accomplished the result. Perhaps without the aid of Mr. Bell, the speaking telephone might not have been brought into public use to this day; but that Drawbaugh produced it there can hardly be a reasonable doubt. We do not question Mr. Bell's merits. He appreciated the importance of the invention, and brought it before the public in such a manner as to attract to it the attention of the scientific world. His professional experience and attainments enabled him to see, at a glance, that it was one of the great discoveries of the century. Drawbaugh was a different sort of man. He did not see it in this halo of light. Had he done so, he would have taken measures to interest other persons with him in it, and to have brought it out to public admiration and use. He was only a plain mechanic; somewhat better instructed than most ordinary mechanics,—a man of more reading; of better intelligence. But he looked upon what he had made more as a curiosity than as a matter of financial, scientific, or public importance. This explains why he did not take more pains to bring it forward to public notice. Another cause of his delay in bringing his invention to public notice was that he was ever indulging the hope of producing speech, at the receiving end of the line, loud and distinct enough to be heard across a room, like the voice of a person speaking in an ordinary tone.

It is perfectly natural for the world to take the part of the man who has already achieved eminence. No patriotic Briton could believe that anybody but Wats could produce an improvement in the steam-engine. This principle of human nature may well explain the relative feeling towards Bell and Drawbaugh in reference to the invention of the telephone. It is regarded as incredible that so great a discovery should have been made by the plain mechanic, and not by the eminent scientist and inventor. Yet the proof amounts to demonstration, from the testimony of Mr. Bell himself, and his assistant, Watson, that he never transmitted an intelligible word through an electrical instrument, nor produced any such instrument that would transmit an intelligible word, until after his patent had been issued; while, for years before, Drawbaugh had talked through his so that words and sentences had again and again been distinctly heard. We do not wish to say a word depreciatory of Mr. Bell. He was original, if not first. He preconceived the principle on which the result must be obtained, by that forecast which is acquired from scientific knowledge, as Leverrier did the place of the unknown planet; but in this, as in the actual production of the thing, he was, according to the great preponderance of the evidence, anticipated by a man of far humbler pretensions. A common astronomer, by carefully sweeping the sky, might have been first in discovering the planet Neptune; while no one but a Leverrier, or an Adams, could have ascertained its existence and position by calculation. So it was with Bell and Drawbaugh. The latter invented the telephone without appreciating the importance and completeness of his invention. Bell subsequently projected it on the basis of scientific inference, and took out a patent for it. But, as our laws do not award a patent to one who was not the first to make an invention, we think that Bell's patent is void by the anticipation of Drawbaugh.

The close vote indicates the difficulty the court experienced in reaching a decision. An article by Dr. I. H. Betz, which appeared in the *Pennsylvania German* magazine (XI [August 1910] 458) states: ". . . the seven justices stood one day 3-4 in favor of Drawbaugh, a day before the final decision which was 3-4 in favor of Bell. . . ."

On behalf of the People's and Overland companies, a petition for rehearing was filed May 7, 1888, on the grounds that the court made certain errors in interpreting the evidence pertaining to: (1) proofs of Drawbaugh's priority, (2) Drawbaugh's conduct, (3) Drawbaugh's ignorance of the date of Bell's invention, (4) Drawbaugh's visit to the Centennial, (5) Drawbaugh's pursuit of his invention, (6) Drawbaugh's neglect to apply for a patent, (7) the tests at New York and Philadelphia, and (8) the construction of the instruments. Also filed with this petition was a full brief with many references to the evidence.

On May 14, Justice Miller delivered the opinion of the court but, as no justice who had united in it asked for a rehearing, the application was denied. Moritz Loth, Simon Wolf, Max Gutmann, Julius Bien, Ernest Marx, *et. al.* then knew the scope and validity of the Bell patent.

IV

The Closing Years

When the telephone litigation opened, the Drawbaughs still were living in the rented house behind the mill. Elcetta wasn't too happy, but there was little she could do but bide her time. Daniel was home only long enough to eat and sleep, for things were moving along at the "old curiosity shop." On October 31, 1882, he received his first patent (No. 266,615) relating to telephones. It covered a carbon transmitter fitted with a double wooden diaphragm, each half of which carried three carbon prisms bearing upon small carbon rollers. The operation was described in this manner: "Resistance offered by said carbons to the passage of an electric current through them will be correspondingly modified rendering said current undulatory and capable of transmitting the character of vocal speech." His attorney, Park Benjamin, expert for the People's Company, had included that magic, vitalizing phrase, "current undulatory."

The papers were issued to Daniel as patentee without assignment. In accordance with his agreement, it is assumed that his interest later was assigned to the People's Telephone Company.

The following year (1883), eight more telephone improvements received patent protection. Included were a signaling system, additional transmitter improvements, and a speaking trumpet for receiving telephones.

Early in the year, Sara J. Haly, a Harrisburg widow of means, had unwillingly taken title to a spacious, single, brick dwelling and outbuildings on four acres of land on "The Hill" at the outskirts of the village. The property adjoined the holdings of the Hertzlers, old friends of the Drawbaughs. Large front windows overlooked the rolling fields; a welcome change from the simple rustic charm of the gristmill's back wall, thought Elcetta. She consulted with the nominal head of the house and reminded him that she had held her peace the year before when he bought the creek workshop from Samuel Hertzler. Now, she wanted her new house. There was some money available and additional funds were coming in from the assignments of the steadily growing list of patents.

Mrs. Haly didn't want the property but had come into ownership by sheriff's deed of April 16, 1883, for settlement of a debt. Since she resided in a lovely home on Front Street, Harrisburg (designed and built by local architect-builder, Stephen Hills, who had erected the state's first capitol buildings at Harrisburg) and was quite satisfied, she would be happy to sell her recently acquired real estate. Daniel took title the following October 9, moved his family forthwith, and returned to inventing.

His happier home environment reflected itself in his work. The year 1884 brought to him thirteen patents on telephonic improvements, and, just to prove that the Drawbaughs really "had it," nephew George took out one also. He devised a transmitter with a gravity carbon block contact and assigned his rights to the People's Telephone Company. The following year, he obtained two more. In 1886, uncle Daniel took his turn again and secured four, and then forgot telephones long enough to complete the neglected, but never forgotten, electromagnetic clock.

That matter was settled August 9, 1887, with patent No. 367,898 which was assigned to Edgar Chellis, the partner whose faith never wavered and whose wallet never was closed.

For months prior to the Supreme Court decision, the impression had been gaining ground, and rumor had it, that Bell would win. But Daniel and his friends turned deaf ears.

Let us, for a moment, consider his status in the case. He had sold his rights, received $5,000 and proportionate share of the stock. All he could hope for was that the stock would become worth more than the paper it was printed on and that the honor of first inventor of the telephone would come to him.

He was not named as one of the defendants; his work had been to witness for the cause to help win the court decision which would make those hopes come true. He had testified for thirty-two working days, answered 1,759 questions, and left the witness chair never to return again, excepting his one appearance on December 13, 1883, to make four brief statements for the Overland case record.

When the bad news came, it left him with mingled feelings of bitter disappointment and fierce resentment which grew into a deep, never-forgotten conviction that he had been betrayed and the secret of his cherished "talking machines" stolen from him. There was little comfort in the knowledge that he then held twenty-six telephone improvement patents when denied the basic one, or that two transmitter interferences had been decided in his favor (Eccard *vs* Drawbaugh, declared July, 1882, decided January, 1886, and Perrin *vs* Drawbaugh, declared August, 1884, decided March, 1885).

His eldest daughter, Laura, had married William Sheely about 1881. Of their four children, Roy, Clifford, Emma, and Edith, the first-born, Roy, alone was living in 1888. Daniel would try to forget and turn his attention to building a house for the Sheelys. A few feet to the west of his home, he erected a comfortable, three-story stone structure.

He would be close to his only grandchild, thinking perhaps the lad, too, possessed mechanical aptitude which could be developed in the coming years. Of course, the workshop fascinated the youngster, as it would have any six-year-old, and he became his grandfather's constant companion. But there were matters still to be settled at Washington.

Protection for certain microphone claims covered by application 126,527 (filed April 3, 1884, as a subdivision of an 1880 application) had been denied Daniel by the Patent Office in 1886. On the grounds of prior public use and the authority of the 1884 Circuit Court opinion which guided the office, it had been decided that he was not the inventor of the matters for which he sought a patent.

On July 17, 1888, Daniel again called up the case. Public use proceedings were instituted by the office and the matter

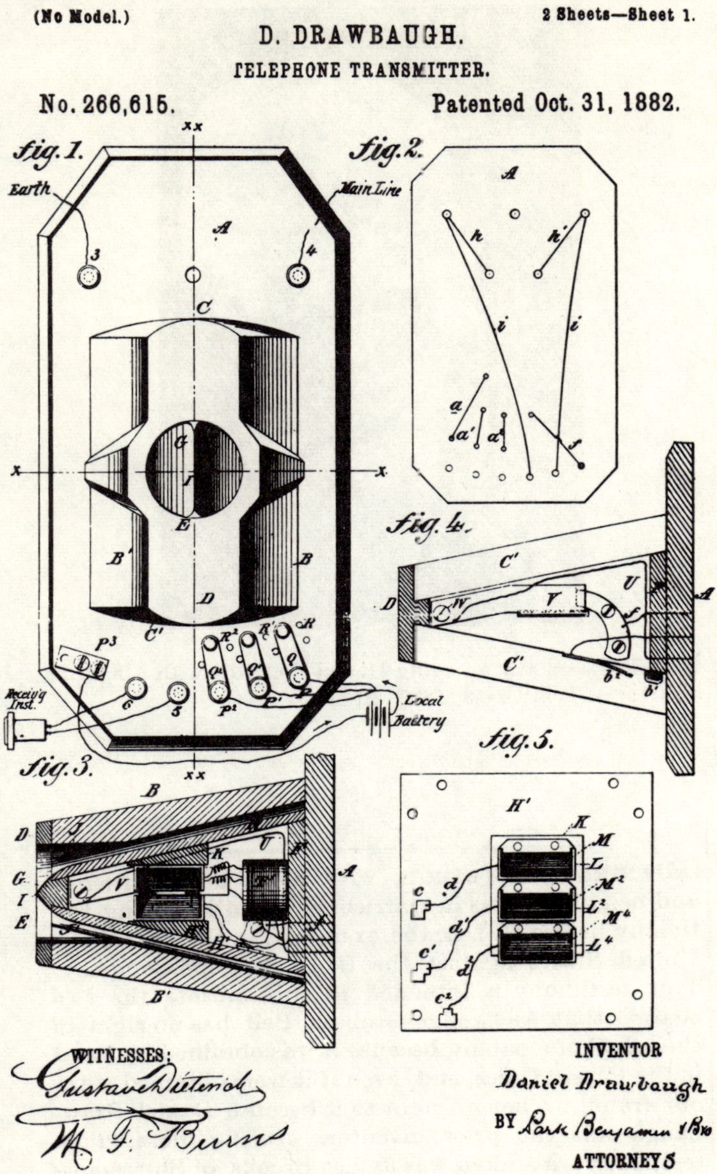

43. The inventor's first telephone patent—improved transmitter of 1882 with double wooden diaphragm, carbon prisms, and rollers.

44, 45. The 1890's saw an aging Daniel—but still with a fighting spirit as this item of February 8, 1890, attests.

Daniel Drawbaugh, with some of his friends and neighbors, was in Harrisburg recently, giving testimony before M. Ott, the examiner, in the suit of the United States against the Bell Telephone Company. This testimony is intended to substantiate the two charges that Alexander Graham Bell has no right to the telephone patent, because it was obtained by fraud in the Patent Office, and, even if it were obtained without fraud, he has no right to it because Daniel Drawbaugh was the prior inventor. Among those whose testimony was taken was James Brooks, of Shiremanstown, Cumberland County, Pa., which is near Drawbaugh's home. He swore that he used the telephone in Drawbaugh's workshop between 1873 and 1874, before Bell had even announced such an invention.

dragged along until the close of 1891 at which time the Commissioner ruled that Daniel was not entitled to a patent.

Then, digging deeper, the claimant resurrected application 19,531 which had been filed October 26, 1880 and denied issue in 1886 and again in 1890. Another appeal to the office in 1893 received a more favorable reception. Assistant Commissioner Fisher granted the petition and the following year patent 514,916, which covered certain features of Daniel's "J" model microphone, issued.

Heartened by this victory the Drawbaugh forces determined to have another go at the Patent Office and, if possible, upset the Commissioner's adverse decision of 1891. Into the Washington Court of Appeals went attorney-partner Lysander Hill, but it didn't take Justice Shephard long to decide that "the Commissioner of Patents did not err in rejecting the (Drawbaugh) application . . . (and) the decision appealed from is therefore affirmed . . ." That bad news came down in May, 1894.

The following year another appeal was taken to the court. Listed were eighteen reasons why the Patent Office should not have adversely decided fourteen other claims. In his brief Daniel stated that he did not expect to be awarded a patent but rather was seeking a decision whereby he would be in position to prove by legal evidence his priority over the other claimants. All that he then desired was to be enabled, by the proper decision, to make the contest.

While this appeal was pending, Senator Cameron, of Harrisburg, Pennsylvania, introduced a bill in the first session of the 54th Congress to grant him relief by authorizing and directing the Commissioner of Patents to grant and issue patents for the applications enumerated in the bill. Pending consideration by a congressional committee, the court quite properly withheld its decision. After two readings the bill was referred to the Committee on Patents and died there.

Then, on September 29, 1896, the court announced its decision. Said Chief Justice Alvey: "Drawbaugh was never the inventor of the transmitter as now claimed by him, and as stated in his affidavits, prior to the dates given in the references cited against him in the patent office . . ." Weary years of litigation again had ended in disappointment. For Daniel the last door had closed.

The creek-side shop was his home and oft-times he said the constant voice of the waters inspired him with a vigor which enabled him to persist in his life's work. Always busy inventing—or fixing, he was never too occupied to spare a few moments for the country lad who wandered in with a broken toy in need of repair, or to provide an explanation of how this or that worked. His pleasing personality gained him hundreds of friends and seldom a day passed without someone visiting the shop to watch and to chat.

A braiding machine was completed and patented in 1889. Next to engage his serious attention were pneumatic tools, a grain weigher, collapsible box, and an electrical generator for the ignition spark of gas engines. For the years 1891 to 1900 inclusive, he received ten patents on the tools, one on the box, one on the grain weigher, nine on telephone improvements, and one on the generator—a total of twenty-two.

His last telephone patent (No. 570,845 issued November 3, 1896) related to substances for telephone electrodes, or more specifically to a platinum substitute for the granular carbon then employed in some transmitters. Others had tried platinum electrodes and discovered the conductivity of the bare metal to be too high. Daniel's process lowered it by electrolytically depositing platinum black on the surface, thus to provide an improved electrode material.

His generator was a radical departure from the then accepted ignition system of batteries and spark coils for gas engines. The writer has been told that for his first experiments he used the old "shocking machine" which had been popular with his youthful shop visitors in the years gone by. Financial help was forthcoming and David Hoffer, of Lebanon, Jacob E. Shettel (the old clock company partner) of Shepherdstown, and Henry B. Eberly (son of Christian Eberly, Daniel's first backer) of Shiremanstown, each held one-quarter interest. The Springfield Engine Company later acquired all right, title, and interest.

The pneumatic tools were of the light, hand type for cutting, chipping, and decorative work, and it has been said that they were used in the construction of the Congressional Library in Washington.

i

The possibility of transmitting telegraphic messages without wires had engaged the attention of scientific and inventive minds for many years. In this country, one of the earlier experimenters was Joseph Henry, after whom the unit of inductance, the Henry, was named, and with whom Bell had consulted while working on his telephone.

Upon his return from England in the fall of 1878, Bell began experiments to devise a wireless telephone. At his Washington workshop he, and his assistant Sumner Tainter, labored with selenium, mirrors and reflectors to solve the problem. Finally, on February 15, 1880, speech was transmitted successfully outdoors, a distance of one-eighth mile, by means of a vibrating beam of sunlight. The "Photophone" was born.

Ways of utilizing the inductive phenomena, which had so plauged the early telephone circuits, were considered by others. Edison devised a telegraphic communications system operated by induction from a moving train to the wires along the right of way. The Phelps' Induction Telegraph Company had similar ideas and by means of an induction coil suspended beneath a car sent the induced currents of a telegraphic message into a cable laid between the tracks. On January 27, 1885, a successful demonstration was given on a section of the New York, New Haven and Hartford Railroad. Other minds considered the problem in wider scope.

In the mountains of West Virginia, Professor Loomis, of Washington, had flown kites attached to copper wires. When at the same height, and in the "natural electric current," telegraphing was "easy and perfect," it was said. Towers had then been erected on hills twenty miles apart, and the professor and his assistant talked successfully between them by means of "aerial telephone." These experiments had been conducted in 1879.

In Italy, Guglielmo Marconi experimented with a system of

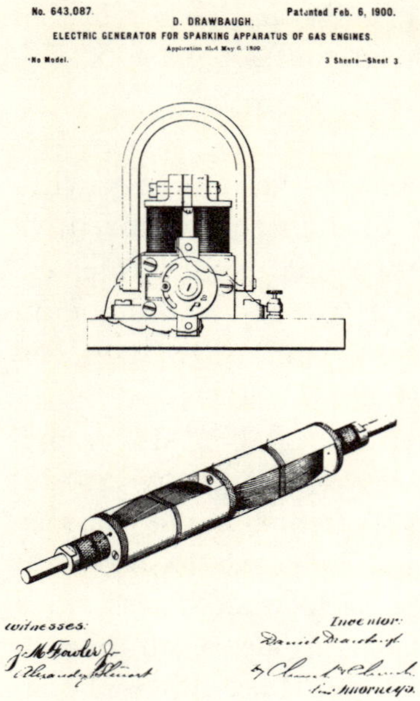

46. The electric sparking generator for gas engines. The twin armature windings at right angles delivered a "continuous and uniform current." Patented 1900.

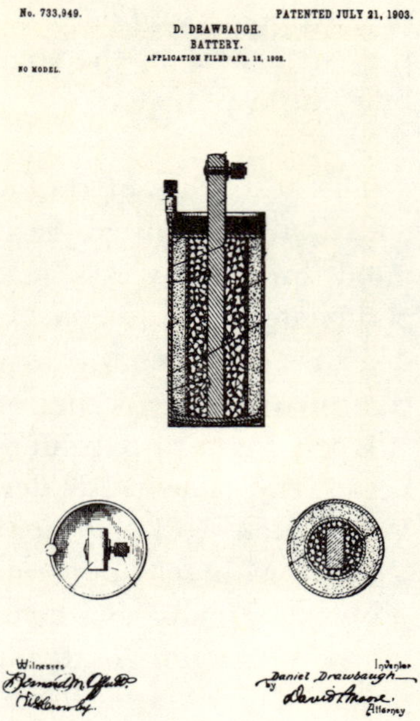

47. Daniel's battery—"cheaper to make – longer life – better efficiency." Patented 1903.

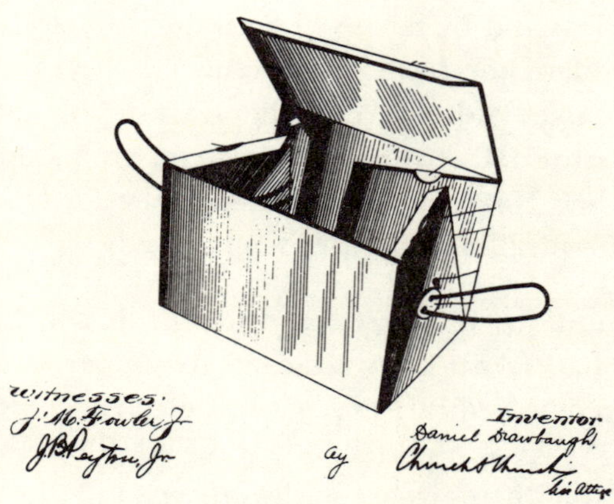

48. The folding lunchbox—a pull on each loop snapped it open to receive the food. Pushing in the ends folded it flat to carry in a pocket. Patented 1901.

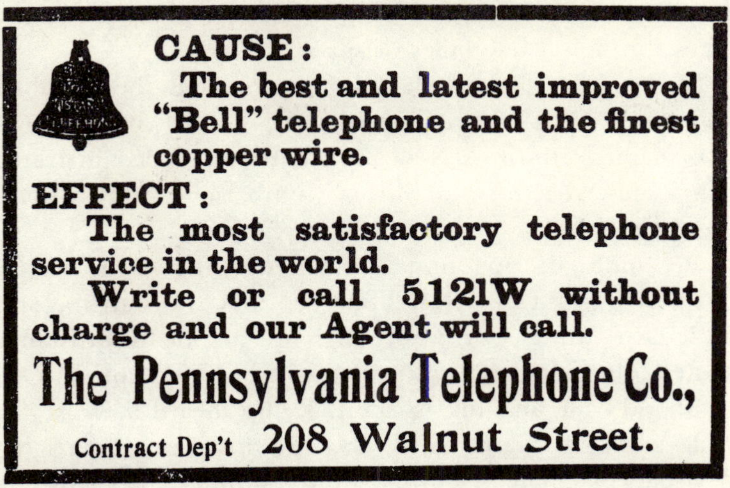

49. In 1893, the first Bell patent expired, in 1894, the second, and independent telephone companies came into being, especially in the rural areas. Through the Midwest, some 6,000 of these little companies were in operation within a few years. It has been said and there is reason to believe that some of Daniel's inventions may have been used by them. Competition was in full bloom with the turn of the century, but his attention then was on the new "wireless."

wireless transmission and in 1896 succeeded in sending messages through a few miles of space.

Inventor Drawbaugh had not overlooked the possibilities and claimed that, as early as 1882, he had, in company with a man named Smith of New York City, succeeded in transmitting a message across the hills of Cumberland County's Doubling Gap.

December, 1901, found Marconi working in an old frame hospital building in New Foundland. A transmitting station at Lizard Point, Cornwall, England, an air-line distance of some 1,740 miles, transmitted signals for his reception. On December 14, he and his assistant, Kemp, heard a weak, Morse code letter "s" signal. This was promptly announced to the newspapers. Some scientific minds doubted; perhaps the signor's ear, or imagination, was working overtime.

To play it safe, the New York City *World* wired Daniel the following day:

MARCÓNI SAYS AT ST JOHNS, N. F. HE RECEIVED THE LETTER "S" IN MORSE CODE WITHOUT WIRES FROM LIZARD, CORNWALL, SEVENTEEN HUNDRED MILES. SIGNAL WAS RECEIVED OVER TELEPHONE FROM KITE AND NOT ON RECORDER. WILL YOU KINDLY TELEGRAPH US AT OUR EXPENSE WHETHER YOU THINK STATEMENT SHOULD BE ACCEPTED WITHOUT QUALIFICATIONS OR WHETHER IT POSSIBLE MARCONI'S EAR DECEIVED HIM. WHAT YOU THINK OF FEASIBILITY OF COMMUNICATIONS ACROSS OCEAN WITHOUT WIRES.

[signed] THE WORLD

Replied Daniel:

YOUR TELEGRAM RECEIVED. IN MY OPINION THE SYSTEM USED BY MARCONI IS FEASIBLE, WIRELESS, AERIAL, OR SO-CALLED ETHEREAL TRANSMISSION IS ACCEPTED BY SCIENTISTS. I AM NOT ABLE TO SAY WHAT LIMIT OF DISTANCE AT WHICH IT MAY BE USED. THE REPORT MAY HAVE BEEN EXAGGERATED, OR SIGNOR MARCONI'S EAR MAY HAVE DECEIVED HIM BY THE STATIC INDUCTION OF THE TELEPHONE RECEIVER.

[signed] DAN DRAWBAUGH

Further successful reception proved both the imagination and the ear of Marconi to be trustworthy and on December 21, 1902, the first wireless message spanned the Atlantic Ocean. Its originator had his patent and the world a new system of communication.

Daniel, too, had been experimenting with a wireless method

but held little faith in the etheric wave transmission theory with its required high, ball-topped poles. Said he: "By experimenting I gradually discovered that the etheric wave had little to do with the transmission of my messages. One day I found that the ground alone carried them . . . I've practically discovered that the air plays no part in the matter. . . ." Right or wrong that was his opinion and in 1901, he had experimented successfully over short distances by his method. The creek workshop was the transmitting station; receiving apparatus had been set up at the Hertzler farm, about half a mile away, and the signals received there. A small boat operating on the Yellow Breeches Creek, half a mile from the shop, also had received successfully. By 1903, voice transmission over a longer range was an accomplished fact. That May, a news representative who interviewed the inventor wrote:

> Mr. Drawbaugh took the reporter into a field along the Yellow Breeches Creek and returned to his workshop and spoke to him over his wireless telephone. There were no connections between the two places. In the field an iron rod was driven into the ground, to which one wire of the telephone receiver was connected. The other wire of the receiver was attached to a wire which ran into the water of the creek about 50 feet away, and which was connected to a tin plate. A current was established by the magneto machine and then Mr. Drawbaugh sent a message over the intervening space. Words were just as clear as over the comercial 'phone of the city and the inventor carried on quite a little conversation . . . When the tin plate was lifted from the water the connection was broken, yet by standing on the bank of the stream and simply touching the plate, the inventor restored it. . . .

A large magneto machine utilizing fourteen pounds of wire for the coils was under construction at the time. This, when completed, would increase the range considerably.

Another news account spoke of the formation of a $3,000,000 company, The United States Wireless Signal Company, organized to market his wireless telephone and telegraph. It was stated that Daniel was to receive $10,000 in cash and a one-tenth interest in the company when his patents were granted. Betz, in his *Pennsylvania German* magazine article of 1910, says that he was offered $25,000 and received a down-payment

50. One of the inventor's first "wireless" receiving stations (1901)—the Hertzler farm on the Lisburn Road. Call letter? Probably "MOO!"

51. Visitors were many at the "old curiosity shop" and members of the Harrisburg Wheel Club especially were welcome. Frank Menath and the "wizard" talk about his latest wonder, the "wireless."

of $400. The company's representative, Doctor W. G. Ward, a former instructor of science and physics, was living at Eberly's Mills, assisting with the experiments, keeping the "wizard" supplied with a never ending stream of "Lancaster County Havanas" (the doctor smoked a pipe), and keeping his eye on developments. The inventor's grandson, Roy, then twenty-one, was living with the family and assisting in the work.

Despite his friendliness and host of friends, there apparently were some who had ill feelings towards Daniel. At least, this is indicated by two mysterious attacks.

In May, 1903, the old inventor was the victim of an attempt at what was said to have been robbery or murder. About one in the morning, Mrs. Drawbaugh was awakened by noises in the bedroom, and saw a man holding a pillow over the head of her husband who was sleeping in another bed. The intruder presumably was administering an anesthetic. She leaped up, aroused two daughters and the grandson and, from the balcony, spread an alarm. The assailant dashed from the house and, joined by two waiting companions, fled down the road toward the village. Shots fired after the fleeing trio missed their mark. The unconscious inventor was revived but showed signs of being drugged. The rooms of the house were in disorder, with valuables lying about, but nothing had been taken.

Two days later an attempt was made to burn the house adjoining the inventor's home. Concerning this, he stated:

> I did not know I had an enemy anymore. Even after my life was attempted on Monday morning, I didn't want to believe it. but it was clear that the men that entered my house did not come to rob me, for they left watches and other valuables practically untouched. I can only think that they entered my home to look for the plans of a wireless telephone which I am now perfecting . . . I have been working on the "wireless" as the papers like to call it, for some time and have had communications running between my house and the shop for months. The shop has not been entered as yet, but you can't tell what will happen. This morning someone set fire to a house of mine next door to our place. The house is of stone and until a month ago was occupied by my son, Charles, who is now in Waynesboro. . . .

It was thought that the idea of the incendiaries was to burn

VOL. LXXII. NO. 205. **Weather: Fair**

MR. DRAWBAUGH BEATS MARCONI

Has Patents for Conducting "Wireless" System Through Ground Instead of the Air---The Old Inventor is Making New and Valuable Discoveries

Professor Ward, who is paid to keep in touch with the progress that Daniel Drawbaugh, the Cumberland county genius of a hundred inventions, makes in wireless telephony and telegraphy, who furnishes the

"It was back in the eighties that I began .experimenting with the wireless. I, too, then did like Marconi, sent my messages from heights. I used the hills, nearby and short poles. He uses poles and complicated appa-

GOVERNMENT HAS EYE ON DRAWBAUGH

Wireless Telephone Could be Used to Great Advantage on Fleets in Time of War---Description of the Wizard's Great-

MILLIONS BACK OF DANIEL DRAWBAUGH

Big Company Formed to Market His Wireless Telephone

52. Newspaper items about Drawbaugh and the "wireless."

TRIED TO MURDER OLD INVENTOR

Daniel Drawbaugh's Wife Found Man Drugging Him While He Slept

Special to "The Telegraph."

Shiremanstown, May 5.—To the wakefulness and courage of his wife alone does Daniel Drawbaugh, the famous old Cumberland county inventor, his life to-day. About one ... ing Mrs. Drawbaugh ... s noises in ...

VOL. LXXII. NO. 10

Weather: Fair.

GUARDS AROUND THE DRAWBAUGH HOME

ld Inventor Works Steadily on Wireless Telephone System---Can Use it Effectively a Mile---Mysterious Visitor From S. Signal Corps.

CONSPIRACY AGAINST DANIEL DRAWBAUGH

Attempt to Burn House of Famous Old Inventor

For the first time since his famous fight with the Bell Telephone Company, Daniel Drawbaugh, the Mill Town inventor of the "speaking machine," fears that he has enemies. This morning at 6 o'clock a house owned by Mr. Drawbaugh, next door to that in which he lives, was set on fire by incendiaries and the aged man has come to the belief, against his will, that there is a conspiracy against him. On Monday morning unknown men entered his bed room and tried to suffocate him.

Mr. Drawbaugh came to Harrisburg ...

call it, for some time and have had communications running between my house and the shop for months. It is a handy way to tell 'em at the house I'm coming home for dinner. When I reach the house the things are on the table.

"The shop has not been entered as yet, so far as I know, but you can't tell what will happen. This morning some one set fire to a house of mine next door to our place. The house is of stone and until a month ago was occupied by my son, Charles, who is now in Waynesboro. The furniture ...

No Company Will Insure Drawbaugh's Property

After a day spent in this city, yesterday, Daniel Drawbaugh, the Cumberland county inventor of the telephone, was compelled to return to his home last evening without getting ...

policies on the place but some months ago Mr. Drawbaugh neglected to continue his payments and the papers were annulled. He visited the company with which he used to conduct business, but it would not take ...

53. Newspaper accounts of attempts at Drawbaugh's life.

the empty house and by so doing set fire to the nearby house of the old inventor and burn that, too. The attempt was thwarted by the "Fourteenth Regiment," the village bucket brigade of ten young men who responded promptly to the early-morning shouts of "Fire!" Drawbaugh's friends were fearful his life was in danger, and guards were employed to watch the shop and the houses. His fire insurance on the properties had been allowed to lapse and after the burning attempt he visited Harrisburg in an effort to have the policies reinstated. No company would accept the risks, but it seems that the presence of watchmen discouraged any further attempts.

The aged inventor obtained an additional patent in 1901 on a folding lunchbox, and in 1903 secured papers for an improved damp-packed cell battery for portable use, an agitator for milk sample tests, and a high-tension magneto machine for wireless use.

On paper, this machine posed a threat to equipment then in use for wireless transmission. In his specification, Drawbaugh declared:

> This invention has reference to a magneto-electric machine; and the object of my invention is the provision of a device which will dispense with the Rhumkorff and other coils and also batteries, which necessarily have to be used in connection with the coils. To this end the invention consists of a magneto device which generates electricity of a very high potential, as is required of coils and all other induction devices, and a device which is novel in construction and combination of parts, as herein set forth.

When, in February, 1904, two patents were granted for his wireless system, the whole secret was revealed.

As has been said, Daniel didn't hold too much faith in the etheric wave transmission theory. He claimed:

> My invention relates to the art of transmitting intelligence, including . . . speech, through the natural media, and depends on my discovery of the practicability of impressing thereon, high-potential currents of electricity, to thereby diminish its resistance to lower potential signaling currents. . . . It comprises a new and improved method of terragraphic and terraphonic transmission,

having as its essential feature the reduction of the resistance of the earth or other natural media between the transmitting and receiving station . . . by impressing on the media at the transmitting station, currents of high potential. These currents may be produced by a magneto-generator of peculiar construction [his magneto-electric machine just referred to]. This phenomenon I believe to be due to a species of earth polarization somewhat similar to the electrolytic polarization. . . . I have found that the transmission of speech is greatly improved by first impressing on the earth these high-potential currents and that with a given battery power of the transmitter, speech becomes perfectly clear and audible. . . .

He had discovered that, after an application of his "polarizing" high-tension current to the earth for a period of two to three minutes, the magneto-generator could be disconnected. Speech transmitted through the earth then could be heard clearly at the receiving station four miles away. He had talked every day for three weeks without the necessity of again "polarizing" the earth!

Small wonder indeed if the earthworms had departed the shop area for healthier climes; after all, those periodic, high-tension jolts were not for *Lumbricus terrestris*—the fisherman's best friend.

ii

After his arduous pioneering in the realm of electronics, Daniel relaxed by devising an improvement on a window blind. A patent came through for it on May 26, 1908, and on June 2 of that year the rights were offered for sale through the Keystone Law and Patent Company of Philadelphia.

He had also devised a process for making small blocks of stove and furnace fuel from coal dust and other ingredients. The product was manufactured at nearby Bowmansdale. From the Harrisburg city directory, it was learned that the Drawbaugh Artificial Fuel Company, Samuel P. Eby, secretary-treasurer, existed in the Union Trust Company Building for at least two years—1909-10.

After the death of Laura Drawbaugh Sheely in 1892, the stone house adjoining the inventor's residence stood vacant. Later, when son Charles married Mabel Hart, they occupied it.

No. 733,948. PATENTED JULY 21, 1903.

D. DRAWBAUGH.

MAGNETO ELECTRIC MACHINE.

APPLICATION FILED MAR. 4, 1902. RENEWED JAN. 12, 1903.

NO MODEL. 2 SHEETS—SHEET 1.

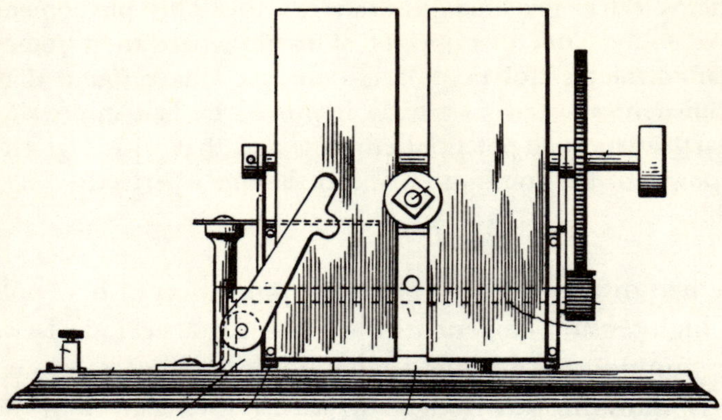

54. The mysterious magnetoelectric machine—it made the "wireless" possible—but the earthworms unhappy.

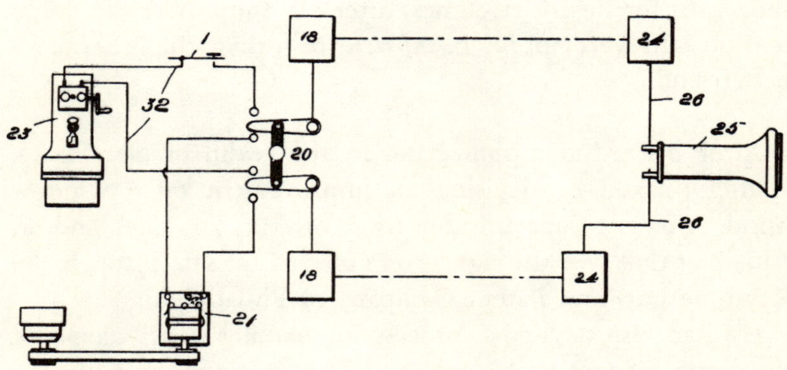

55. The system—this schematic diagram revealed all. Magnetoelectric machine (21) was operated by its drive-motor. Metal plates (18) were buried in the ground at the transmitter. The double-pole, double-throw switch (20) connected the ground plates with the machine when in "down" position. After several minutes of "conditioning" the earth, switch (20) was thrown to "up" position, cutting out the machine and cutting in the telephone instrument (23). Speech was transmitted through circuit wires (32) to ground plates (18) and thence through the earth to plates (24) at the receiving station. Sounds would be heard from receiver (25).

His work took him away a good bit of the time, his wife stayed with her folks, and the couple seldom were there to enjoy the house. For Daniel and Elcetta, things just weren't the same anymore, so they decided to make their last move.

Daughter Iola, and her husband, Hal Oyster, and the children were living on Cumberland Road (now South Seventeenth Street), Camp Hill. Their spacious stone and frame house would be just the thing for the old inventor and his family, and Elcetta would have more room for the piano. Next door was a double cement block house, also owned by the Oysters into which they could move and be close to the old folks.

Doctor J. Markwood Peters, a physician of Steelton, was interested in buying the inventor's two Eberly's Mills houses. On March 31, 1904, he took title and on the same day Daniel settled for the Cumberland Road property. The Drawbaughs moved to their new home, but Daniel still retained his beloved old shop at the creek. Each morning, the aged inventor walked the mile and a half to work, and each evening home again—for seven years.

His last two patents dealt with something which always seemed to trouble him a great deal—money. In 1910, at the behest of Robert L. Myers of the Cumberland Valley Bank (then soon to become the Lemoyne Trust Company), he set to work to devise a coin assorter to expedite the work of the tellers. Soon, a model was completed. The coins were dumped into a hopper, the handle turned, and each denomination fell into place in its proper tube. A supplementary device was planned to arrange the packaging of the coins into rolls of predetermined amounts.

At this time, an account of an interview given by the old inventor at his creek workshop appeared in the June 16 (1910) issue of the Harrisburg *Star Independent,* over the initials of staff correspondent, B. F. M., Jr. (probably Benjamin F. Myers, Jr., son of the president of the publishing company). He wrote of the coin assorter, of a new metal discovered which baffled the analytical efforts of the chemists who had examined it, of a rotary engine to replace the piston type, an account of $3,500 received by Daniel from the com-

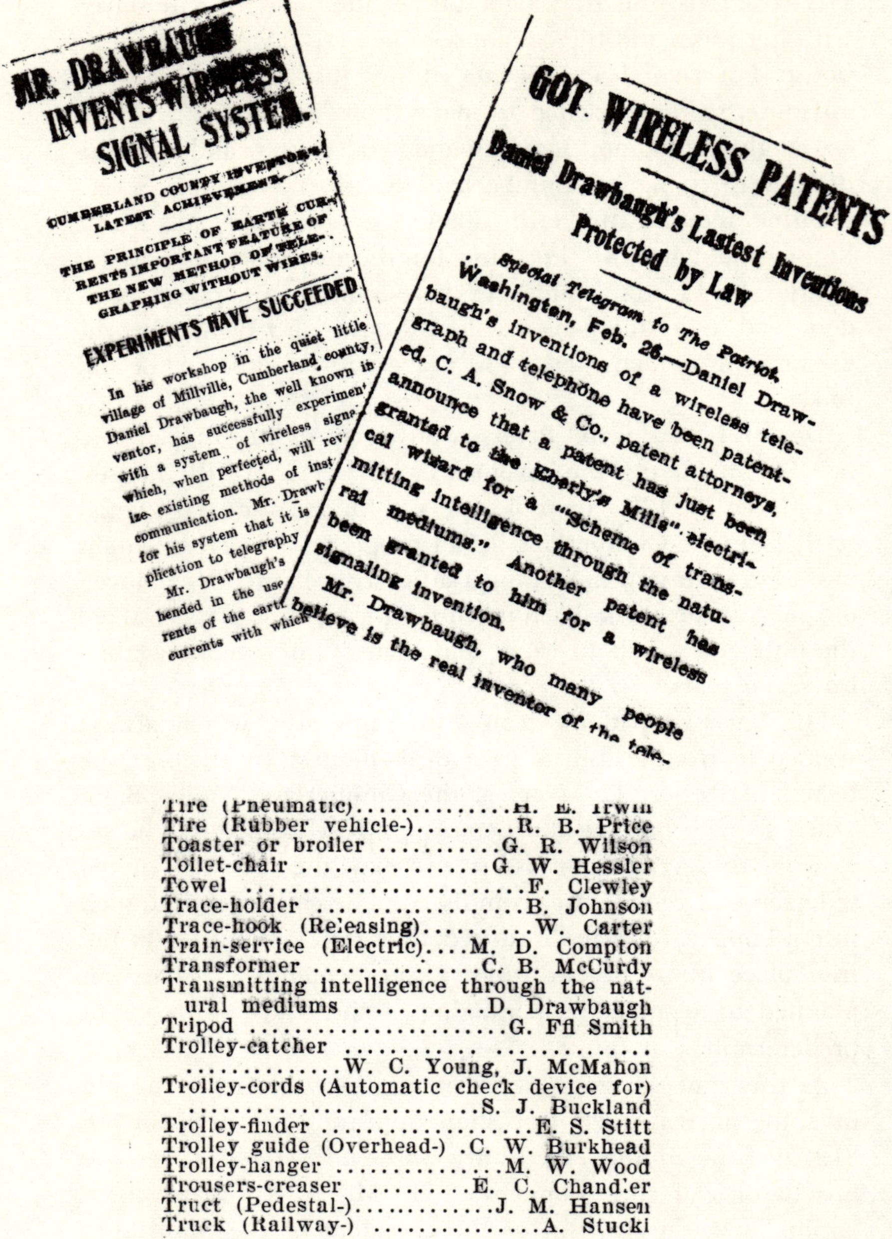

MR. DRAWBAUGH INVENTS WIRELESS SIGNAL SYSTEM.

CUMBERLAND COUNTY INVENTOR'S LATEST ACHIEVEMENT.

THE PRINCIPLE OF EARTH CURRENTS IMPORTANT FEATURE OF THE NEW METHOD OF TELEGRAPHING WITHOUT WIRES.

EXPERIMENTS HAVE SUCCEEDED

In his workshop in the quiet little village of Millville, Cumberland county, Daniel Drawbaugh, the well known inventor, has successfully experiment with a system of wireless signa which, when perfected, will rev ize existing methods of inst communication. Mr. Drawb for his system that it is plication to telegraphy Mr. Drawbaugh's hended in the use rents of the eart currents with which

GOT WIRELESS PATENTS

Daniel Drawbaugh's Lastest Inventions Protected by Law

Special Telegram to The Patriot.

Washington, Feb. 26.—Daniel Drawbaugh's inventions of a wireless telegraph and telephone have been patented. C. A. Snow & Co., patent attorneys, announce that a patent has just been granted to the Eberly's Mills' electrical wizard for a "Scheme of transmitting intelligence through the natural mediums." Another patent has been granted to him for a wireless signaling invention.

Mr. Drawbaugh, who many people believe is the real inventor of the tele-

Tire (Pneumatic)..............H. E. Irwin
Tire (Rubber vehicle-).........R. B. Price
Toaster or broiler ....G. R. Wilson
Toilet-chair ....G. W. Hessler
Towel ....F. Clewley
Trace-holder ....B. Johnson
Trace-hook (Releasing)..........W. Carter
Train-service (Electric)....M. D. Compton
Transformer ....C. B. McCurdy
Transmitting intelligence through the natural mediums ....D. Drawbaugh
Tripod ....G. Fl Smith
Trolley-catcher ....W. C. Young, J. McMahon
Trolley-cords (Automatic check device for) ....S. J. Buckland
Trolley-finder ....E. S. Stitt
Trolley guide (Overhead-) .C. W. Burkhead
Trolley-hanger ....M. W. Wood
Trousers-creaser ....E. C. Chandler
Truct (Pedestal-)............J. M. Hansen
Truck (Railway-) ....A. Stucki

56. The technological march of 1904—along with a pneumatic tire and a trousers-creaser came a toilet chair—for the infant "wireless?"

pany organized to market his pneumatic tools was given, as was the only explanation your writer discovered of Daniel's oft repeated assertion that his telephone secret had been stolen from him. Wrote the staff correspondent, "Under the guise of friendship, a man wormed his way into the inventor's confidence. The secret learned, he sold it to another."

Reference also was made to a bill in Congress which was to give him one dollar a year for every telephone in use, but a search of the Congressional Record Index of the Sixty-first Congress (first, second, third sessions, 1909-11) by the documents section of the Pennsylvania State Library and the writer failed to reveal its existence.

Billy Bowman, venerable retired schoolteacher and former member of the Pennsylvania State Assembly, now eighty-seven years young, then was doing part-time work at the shop. To him were relegated the tasks of reconditioning unruly pneumatic tools, and the construction of the coin assorter. Billy was well pleased with the completed model and wanted a half interest in it. The transfer was made officially under the seal of the Patent Office on December 22, 1910—Billy's Christmas present. While that particular model was not a great success commercially, Billy still cherishes it highly—as well he should—as the "wizard's" last patent, and the last to be made in the old shop.

It marked the end of an era; the final curtain had descended on the "old curiosity shop."

Many times, a possible solution to a problem suggested itself to the inventor in the middle of the night. As was his lifetime habit, he would rise, dress, and walk to the shop to try it regardless of the hour—for the matter just couldn't wait until morning. No one was surprised to see Daniel on the streets or out along the Milltown Road at any hour. Now, that was a thing of the past. In July, 1911, the old machine shop was sold to Congressman-attorney Marlin E. Olmsted, owner of the adjoining land. A small workshop was obtained on Market Street near the inventor's home. He finished his compressed air machine devised to clean the outside of cars, and then spent his time sharpening tools, repairing bicycles, doing odd jobs of work, and, of course, experimenting with new devices which came to mind.

U. S. Wireless Signal Company representative, Doctor Ward, completed his Eberly's Mills mission and bowed out in a blaze of glory with his newly acquired bride, Maude Drawbaugh, who resigned her position as bookkeeper with a Harrisburg firm to move to New Jersey with her husband. Grandson Roy, too, had married and was living on the Trindle Road on the outskirts of Camp Hill.

Elcetta and Daniel were in the twilight years. Now, only they, daughter Belle, and son Charles, who occasionally came home, were left to occupy the large house. But the old inventor kept himself too busy to be lonely.

Thursday, November 2, 1911, was a clear nippy day, just the sort of weather to walk to Daniel's shop and see what was new.

The word had gotten around that the "wizard," with the help of Charles, had just put the finishing touches to a new device—something really different this time—a wireless burglar alarm! All morning long, friends and the just curious were in and out of the Market Street shop. Everyone wanted to see it work, and then wondered how it did—a dozen demonstrations, a hundred questions followed by a hundred answers. Over in Harrisburg, the newspapers received word of the "open house" and around noon a couple of newsmen dropped in. "The Edison of the Cumberland Valley," as he was sometimes referred to in print, was good copy. As he recounted his achievements and told of the manner in which, as he always maintained, his telephone secret had been stolen by a "spy," he grew visibly excited, stammered, laid down an instrument which had just been handed to him, and dropped into a nearby chair.

Doctor Peters happened in a few minutes later and found the old inventor in what he supposed to be an attack of vertigo. Ordered home, he was placed in bed. Throughout the day and evening, his condition gradually grew worse. Doctors Peters and Bowman were called in at midnight. About a quarter to three, he feebly expressed a desire to talk to Charles, but when the son leaned over the bedside—there were no

57. Pocket size pipe organ—that's what the writer thought until Billy explained the intricacies of the coin assorter—the "wizard's" last patent.

8—303.

DEPARTMENT OF THE INTERIOR.

UNITED STATES PATENT OFFICE.

RECEIVED and RECORDED on the 28th *day of* December, 1910, *in Liber* Y 85, *page* 285 *of Transfers of Patents.*

IN TESTIMONY WHEREOF, I have caused the seal of the Patent Office to be hereunto affixed.

Exd Mad E. B. Moore.
Commissioner of Patents.

8—1897

......, of Camp Hill, County of Cumberland, State of Pennsylvania have invented certain new and useful improvements in Coin Assorters for which I have made application for Letters Patent of the United States;

And Whereas, William C. Bowman of Lemonye, County of Cumberland, State of Pennsylvania, is desirous of acquiring an interest in said invention and in the Letters Patent to be obtained therefor:

Now, Therefore, to all whom it may concern, be it known that for the sum of One Dollar, and other valuable considerations, the receipt of all of which is hereby acknowledged, I have sold, assigned and transferred, and by these presents do sell, assign and transfer, unto the said William C. Bowman one-half of the entire right, title and interest in and to the said invention, as fully set forth and described in the specification incorporated in application for United States Letters Patent Serial Number 596,806 Filed December 12, 1910.

And I hereby authorize and request the Commissioner of Patents to issue the said Letters Patent to the said William C. Bowman and myself in accordance with this assignment.

In testimony whereof I have hereunto set my hand and affixed my seal this 22 day of December 1910.

Signed in presence of

Maude Drawbaugh
John W. Bowman

Daniel Drawbaugh [SEAL]

[SEAL]

58. Billy's Christmas present—"I have sold, assigned and transferred . . . unto the said William C. Bowman, one-half of the entire right, title and interest in and to the said invention [the coin assorter] . . . ."

words. Daniel Drawbaugh had passed away as he had chosen to live—peacefully.

On a raw, gloomy, Monday afternoon, in a cold, driving rain, his body was lowered into its last resting place in St. John's Cemetery.

Now, the statement has been made, but I couldn't prove its truth, that of the flowers at the service, the nicest piece was a large, bell-shaped wreath.

Daniel Drawbaugh died intestate. Letters of administration were issued November 13, 1911, to his son Charles, who had been appointed administrator. However, the new motorcar industry beckoned him on to Detroit, the letters of administration were vacated, and the Lemoyne Trust Company appointed in his place, December 12, 1912.

The stone house was sold to Marlin E. Olmsted for $3,725, and the adjoining double dwelling was purchased by William S. Snyder for $3,850. To liquidate the principal outstanding debts, a $2,343 mortgage was satisfied and notes totaling $4,024 were paid, as were attorney's fees, burial, and miscellaneous expenses.

Long years before, Daniel Drawbaugh had been quoted as saying that before he would give up working on his telephones, he would go to the poorhouse. Fortunately, that didn't come to pass.

In payment for a lifetime of struggle, bitter disappointment —and tears, he was able to leave to his widow, cash and personal property of almost $350.

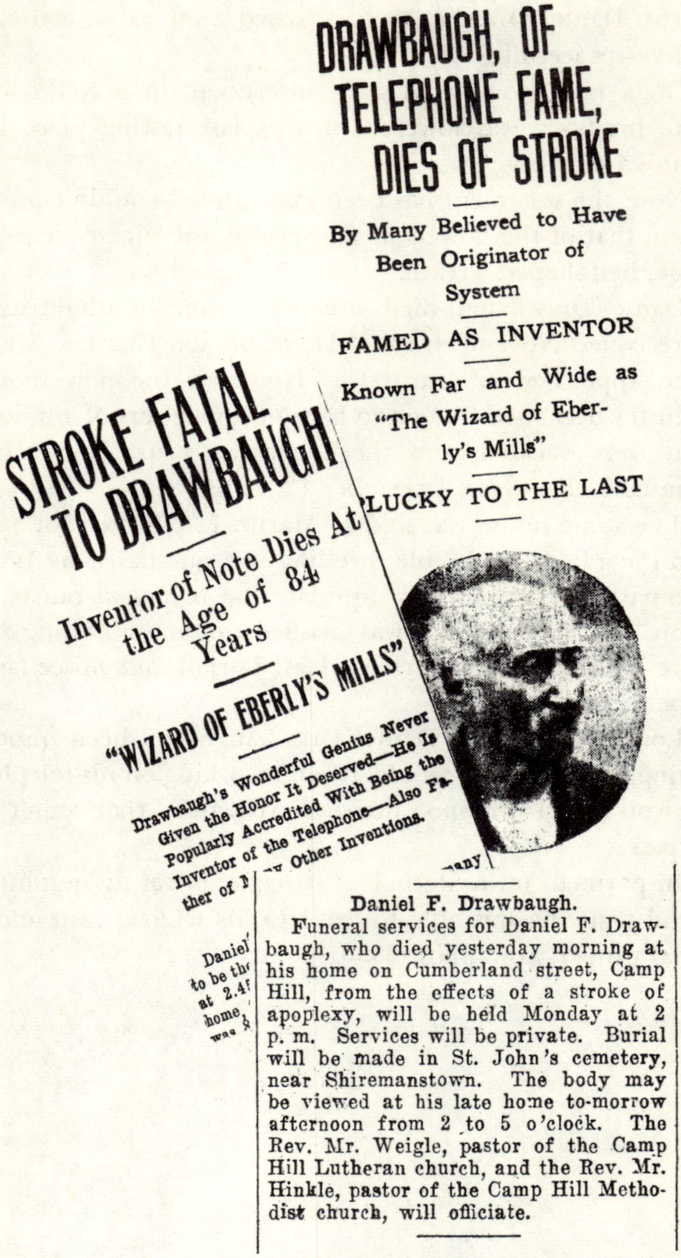

DRAWBAUGH, OF TELEPHONE FAME, DIES OF STROKE

By Many Believed to Have Been Originator of System

FAMED AS INVENTOR

Known Far and Wide as "The Wizard of Eberly's Mills"

PLUCKY TO THE LAST

STROKE FATAL TO DRAWBAUGH

Inventor of Note Dies At the Age of 84 Years

"WIZARD OF EBERLY'S MILLS"

Drawbaugh's Wonderful Genius Never Given the Honor It Deserved—He Is Popularly Accredited With Being the Inventor of the Telephone—Also Father of Many Other Inventions.

Daniel to be the at 2.4 home was

Daniel F. Drawbaugh.

Funeral services for Daniel F. Drawbaugh, who died yesterday morning at his home on Cumberland street, Camp Hill, from the effects of a stroke of apoplexy, will be held Monday at 2 p. m. Services will be private. Burial will be made in St. John's cemetery, near Shiremanstown. The body may be viewed at his late home to-morrow afternoon from 2 to 5 o'clock. The Rev. Mr. Weigle, pastor of the Camp Hill Lutheran church, and the Rev. Mr. Hinkle, pastor of the Camp Hill Methodist church, will officiate.

59. Newspaper clippings of Drawbaugh's death.

60. The Drawbaugh memorial in St. John's Cemetery. Here are buried son Harry, who died Christmas Day, 1881, at the age of 14 years, 2 months, and 7 days; Daniel, the father, who died at the age of 84 years, 3 months, and 19 days; Elcetta, the mother, who survived her husband by 1 year, 6 months, and 4 days, having died on May 7, 1913, at the age of 79 years, 5 months, and 6 days; and the last surviving child, Belle, who died circa April 24, 1953, at the age of 85 years.

Appendix C. The Real Estate Holdings of Daniel Drawbaugh

61*a*. His first realty venture—the double house at Eberly's Mills (showing the rebuilt blacksmith shop) as it appears today. Acquired from Elisha and Hannah Hauck, March 13, 1868. Consideration $2,310. Sold to Daniel Fettrow, April 1, 1876, for $2,200.

61*b*. East Main Street, Mechanicsburg – purchased from Eli Wise, April 1, 1876. Consideration $1,740. Here the Drawbaughs lived for one year (April, 1876-77). Conveyed to Joseph Bricker, December 4, 1896, for a consideration of $1,200.

61*c*. The old shop—Daniel's hugger-mugger purchase of about 1882, backfired later when he lost his unrecorded deed. On June 28, 1911, his old friend, Samuel M. Hertzler, then a widower, executed a new deed. No consideration given. On July 11, of the same year, the property, with its 2 acres, 39 perches, was conveyed to M. E. Olmsted for $850.

61*d*. His third holding at Eberly's Mills was his home from 1883-1904. Acquired from Sara J. Haly, October 9, 1883. The brick house and outbuildings were situate upon a plot of 4 acres, 19 perches. Consideration $3,000. From this parcel, one lot was conveyed to John S. Hamilton, December 8, 1894. No consideration given.

61*e*. The home of Laura Drawbaugh Sheely—adjoining the brick house of Daniel Drawbaugh, it had been erected by him in the late 1880's. About the turn of the century, it was occupied by his son Charles and wife, Mabel Hart Drawbaugh. Set afire by unknown persons in May, 1903.

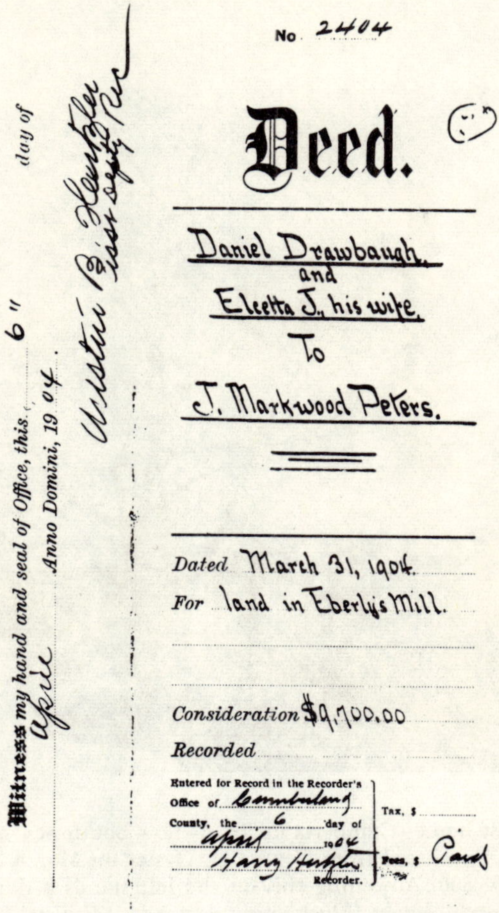
No. 2404

Deed.

Daniel Drawbaugh
and
Elcetta J., his wife.
To
J. Markwood Peters.

Dated March 31, 1904.
For land in Eberly's Mill.

Consideration $9,700.00
Recorded

Entered for Record in the Recorder's Office of Cumberland County, the 6 day of April 1904
Harry Hertzler Recorder.
Tax, $
Fees, $ Paid

Witness my hand and seal of Office, this 6" *day of* April *Anno Domini, 19*04

61*f*. On March 31, 1904, Daniel and Elcetta J. Drawbaugh conveyed the stone and the brick properties to Doctor J. Markwood Peters, for a consideration of $9,700. It is of interest to note that this indenture bears the wife's signature, fully written, as *Elcetta* J. Drawbaugh. She did not appear as a witness in the People's Company case, but her deposition was taken at Harrisburg on March 11, 1885, in the Overland case. As has been mentioned, certain evidence from this case became a part of the People's case record by stipulation. She testified under the name of *Elcinda* Jane Drawbaugh and sealed her deposition with her (x) mark.

61*g*. His last home—Cumberland Road—now South Seventeenth Street, Camp Hill. Purchased from George H. Oyster on March 31, 1904. Consideration $3,800. Adjoining this on the left stands a double, two and one-half story, cement blockhouse acquired October 3, 1905, from G. H. Oyster and wife for $4,500.

Appendix D. THE PATENTS OF DANIEL DRAWBAUGH

Date	Device	Patent Number
1851	Stave Jointing Machine	8,505
1855	Stave Machine	12,900
1863	Millstones	38,296
"	Machine for leveling faces of millstones	38,472
1865	Nail Plate Feeder	51,435
1866	Measuring Faucet	59,792 – 3
1867	Nail Plate Feeding Device	71,148
1879	Earth Battery for Clocks	211,322
1880	Water Motor	225,577
"	Measuring Faucet	229,592
1882	Telephone Improvements	266,615
1883	" "	272,865 – 6
"	" "	276,136 – 7
"	" "	287,111
"	" "	290,978 – 9 – 80
1884	Telephone Improvements	291,311 – 12
"	" "	291,472
"	" "	293,154
"	" "	295,741 – 2
"	" "	297,578 – 9
"	" "	298,676
"	" "	303,627 – 8 – 9
"	" "	307,026
1886	Telephone Improvements	339,745
"	" "	344,818
"	" "	354,541 – 2
1887	Magnetic Clock	367,898
1889	Braiding Machine	412,369
1891	Pneumatic Tool	458,291
"	" "	464,820 – 1
1892	" "	472,495
"	" "	479,061
"	" "	482,942
1893	" "	504,802 – 3
"	Grain Weigher	507,531
Date	Device	Patent Number
1894	Pneumatic Tool	530,934
"	Telephone Improvements	514,916
1895	" "	540,781
"	" "	540,959 – 60 – 61

"	" "	552,469
1896	" "	567,966
"	" "	568,526
"	" "	570,845
1898	Pneumatic Tool	606,961
1899	Collapsible Box	632,451
1900	Electric Generator	643,087
1901	Folding Box	676,393
1903	Magneto	733,948
1903	Battery	733,949
1903	Milk Tester Agitator	747,870
1904	Wireless Transmission	752,533 – 4
1908	Window Blind	888,792
1911	Coin Assorter	989,238
1911	" "	989,522

Bibliography

In the preparation of this work, the following sources were consulted:

American Bell Telephone Company Versus Overland Telephone Company. Printed transcript of the testimony.

American Bell Telephone Company Versus People's Telephone Company. Printed transcript of the testimony.

Biographical Annals of Cumberland County. Jeremiah Zeamer, 1905.

Congressional Record. Index and Vols. 44, 45, 46, covering the first, second, and third sessions of the Sixty-first Congress.

Cumberland Blue Book. Jeremiah Zeamer, historian. J. R. Schwartz, 1908.

History of Cumberland County. Conway Wing, editor. James D. Scott, 1879.

Old Wires and New Waves. Alvin Fay Harlow. Copyright 1936. Material used by permission of the publisher, Appleton-Century-Crofts, Inc.

Pennsylvania German Magazine, Vol. 11, 1910. Express Printing Company, publisher. Quoted by permission of the present copyright owner.

Scientific American, files from 1876 to 1894 inclusive.

Speaking Telephone Interferences A-L and No. 1. Briefs for Alexander Graham Bell and Francis Blake.

U. S. reports covering the Circuit and Supreme Court decisions.

Newspaper articles. Various papers and dates as quoted in the text.

Patent Office Records. Copies of patents pertaining to inventions of Daniel Drawbaugh. *Decisions of the Commissioner of Patents*. 1877 to 1897 inclusive. Personal interviews as listed in the introduction. Records of the Register of Wills and Recorder of Deeds for Cumberland County, Pennsylvania.

Index

R

S

T

U

V

W

Y

Z